DESIGN & DETAIL
Country House
112_전원주택

112_Country House

Publisher & Editor Jeong, Jiseong

www. capress. co. kr
E-mail. capressconcept@gmail.com
capressmaru@gmail.com

Editorial Jounalist Capress Co., Ltd
Design CONCEPT, MARU

Published by CA Press Co., Ltd
© Copyright 2021

Address. A10-051, 550, Misa-daero, Hanam-si
Gyeonggi-do, Korea
Phone. 82-2-455-8043
Management & Sales. 82-2-455-8040
Fax. 82-031-8027-3709

Printed in Seoul. Korea

Price USD 59, Euro 67

112_전원주택

발행 및 편집인 정지성

www. capress. co. kr
E-mail. capressconcept@gmail.com
capressmaru@gmail.com

기획. 취재 CA현대건축사
디자인 월간 컨셉 편집부

발행 CA현대건축사
© Copyright 2021

주소 경기 하남시 미사대로 550
현대지식산업 A10–051
대표전화 82–2–455–8043
관리 영업 82–2–455–8040
팩스 82–031–8027–3709

인쇄 대한민국. 서울

정가 64,000원

* 이 책에 게재된 기사나 사진의 무단 복제 및 전재를 금합니다.

Today, In the era of the establishment of a five-day work week and per capita income of $30,000, the demand for country houses is explosively increasing in Korea. Just 10 years ago, country houses were the exclusive property of some wealthy people in Seoul during the week and in the countryside on the weekends, but these days, with the development of transportation and communication, and the return of the baby boomer generation, rural urban populations are increasing. IIn Design & Detail 112, only unique and outstanding country works were collected and classified according to size, and the architect's design concept and detail were densely edited.

주 5일제 근무제 정착 그리고 1인당 국민소득 3만 달러 시대를 맞은 오늘날 한국에서는 전원주택 수요가 폭발적으로 증가하고 있다. 불과 10여년 전만 해도 전원주택은 주중에는 서울에서, 주말에는 농촌에서 보내는 일부 부유층의 전유물이었지만, 요즈음 교통과 통신의 발달 그리고 베이비붐 세대의 귀향 등으로 농촌 지역으로 이주하는 전원 속의 도시 인구가 늘고 있다. 디자인 & 디테일 112에서는 국내의 독특하고 뛰어난 전원 작품들만을 모아 규모에 따라 분류하고, 건축가의 디자인 컨셉과 디테일을 밀도 있게 편집하였다. —편집자주—

CONTENTS

Country House

전원주택 ~100㎡
Country House ~100㎡

006 One House
영동주택 / DRAWING WORKS

016 Maengdong House
맹동주택 / How Architects Inc.

026 Landschaft
여주 기쁨의 주택 / Architects601

036 Ger House
게르 하우스 / atelierjun

044 Gaonnuri
가온누리 / Gong-Yu Architecture

054 Pause
파우재 / Architectural Design Group HAEDAM

062 Hokepos
호케포스 / NAOI + PARTNERS

072 Oreum Madang House
오름마당집 / JNPeople Architects

전원주택 100~150㎡
Country House 100~150㎡

080 Cheongju Empty & Full House
청주 비담집 / RICHUE Architecture

092 House_B382
하우스_B382 / Moku Architects

100 Oh U House
오유당 / WE Architects

108 Cheongdo Imdang-ri House
청도 임당리 주택 / MOON ARCHITECTS

118 HwaDam Byeol Seo
화담별서 / ilsangarchitects

128 BB4
BB4 / MaroAN Architects

138 The Sky Court House
스카이 코트 하우스
/ Gyeongpiri Architecture Powerhouse

전원주택 150~200㎡
Country House 150~200㎡

146 House on hill by the sea
바닷가 언덕위 집 / JNPeople Architects

154 YUYUJAJEOG
유유자적 / admobe architect

162 Boryeong House
보령 주택 / WELLHOUSE

170 Pyeongdamjae
평담재 / Office for Appropriate Architecture

182 Mok-dong House
목동주택 / JDArchitects

192 Green House
푸른집 / PLANO architects & associates

204 OYEONJAE
오연재 / Zerolimits Architects

전원주택 200㎡~
Country House 200㎡~

214 Sangseon-won
상선원 / We Architects

226 JAEHO HOUSE
재호가 / JNPeople Architects

236 Bi-House
수동 바이-하우스 / WELLHOUSE

246 Kang An Dang
강안당 / Sung Architects Group

One House Yeongdong, Korea

DRAWING WORKS

Architects DRAWING WORKS / Youngbae Kim **Project Team** Suah Hwang **Location** Yeongdong, Korea **Site Area** 670㎡ **Bldg. Area** 38.39㎡ **Gross Floor Area** 59.26㎡ **Bldg. Scale(Floor)** 2FL **Structure** Wood Structure+Reinforced Concrete(Foundation) **Max. Height** 6.7m **Exterior Finish** Monocouche **Construction** Analog Atelier **Client** Jaehoon Lee **Photographer** Jaekyeong Kim

설계 드로잉웍스 / 김영배 **설계참여자** 황수아 **위치** 충청북도 영동군 상촌면 고자리 437 **대지면적** 670㎡ **건축면적** 38.39㎡ **연면적** 59.26㎡ **규모** 지상2층 **구조** 경량목구조+철근콘크리트(기초) **최고높이** 6.7m **외부마감** 모노쿠쉬 **시공** 아날로그 아틀리에 / 류재호 **발주처** 이재훈 **사진** 김재경

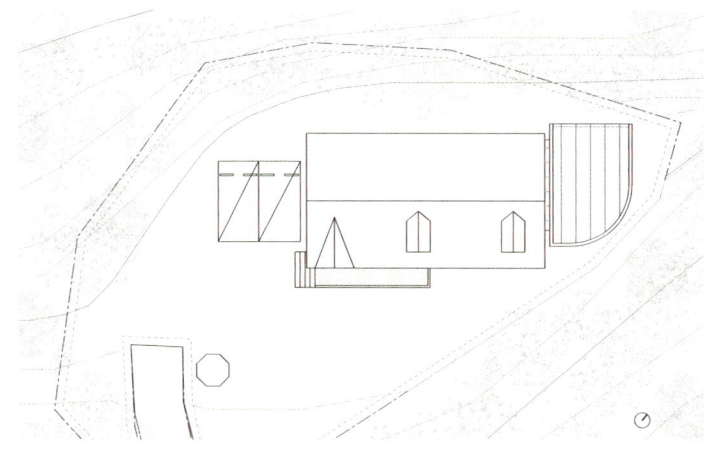

Site Plan

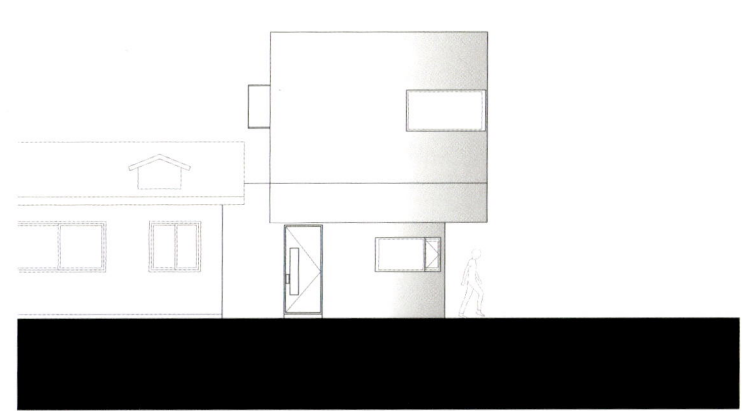

Elevation I

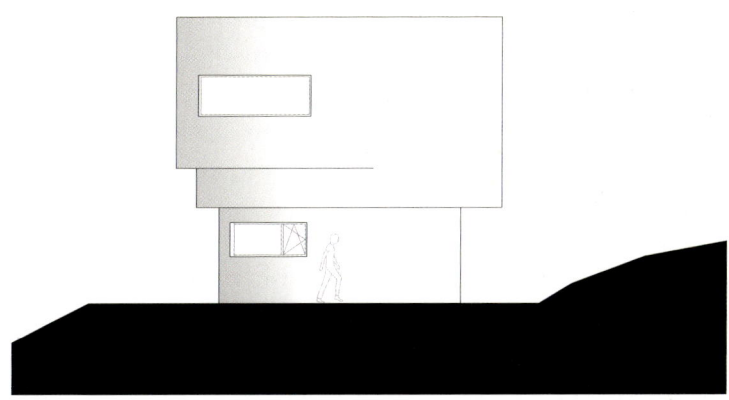

Elevation II

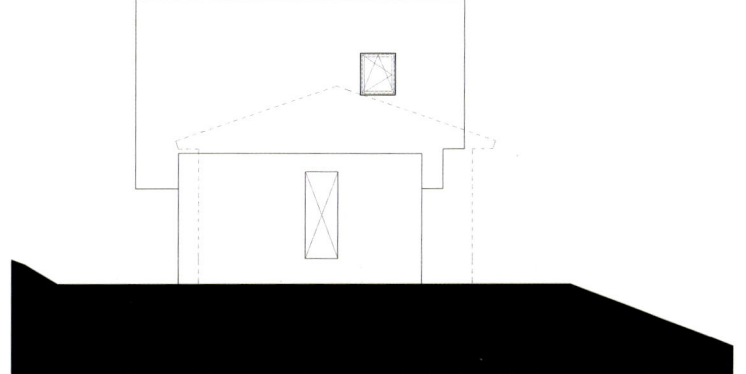

Elevation III

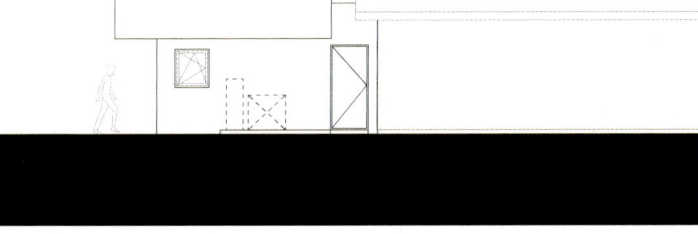

Elevation IV

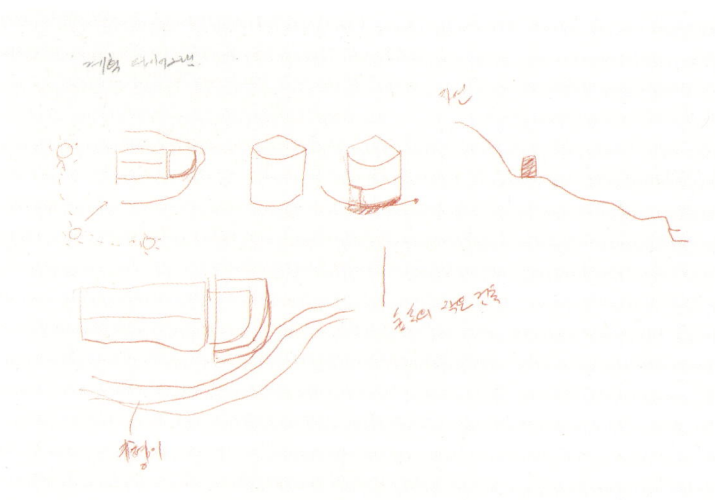
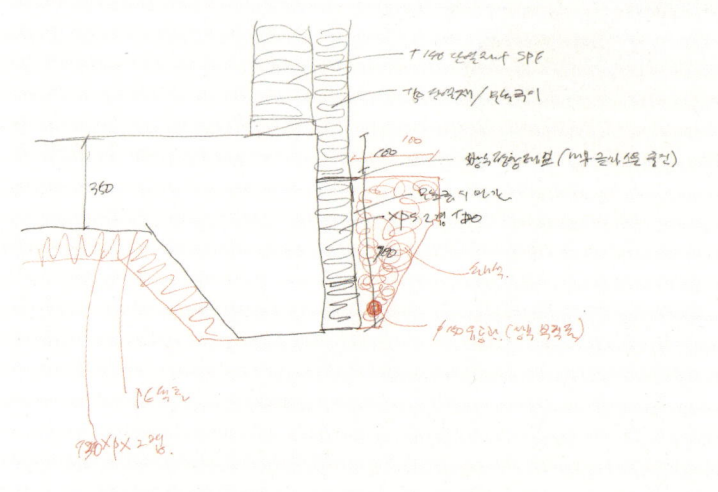

Sketch

A Solo Forest Dwelling

A farmhouse, on a site of an area of 670㎡, sits atop a steep slope at a height of 60m from the access road. The client wanted a house in which to live with his parents so he could move to this remote, tranquil town. He approached us to fulfill the design of a new house in which he might live alone as neighbour to the existing farmhouse, within the confines of the lot owned by his parents. This house had to feature a kitchen and storage space on the first floor, and a room that connected to the existing farming house on second floor. In other words, he wanted to use the first floor as an open, shared space, and the second floor as a private space. Before the client met us, he drew up a floor plan to sketch out the arrangement and size, also studying and compiling data to do with various eco-friendly materials and facilities. By examining these findings, he conceived of his desired construction technique, in partnership with us. Most significantly, we contributed to the detailed drawings, developing six alternatives, sharing a model and 3D modeling data with the client – even though it is a small house, we were aware that we would only have to draw up new floor plans should the furniture position change. In light of the planned spatial arrangement, we set upon devising a new building for the client inside the right side of the existing building. The building could be connected to his parents' home via the kitchen and multi-use room. However, this was close to the perimeter of the lot and because of the outside path the area available to the first floor was only 20m2, and so we designed 1m of the second floor as a cantilever structure to expand the space and manufactured furniture that adapted to the curved shaping of the walls to improve its practicality. At this same stage, we also opened up the possibility of a panoramic view within the space.

A Problematic Site Faced with a Builder's Tenacity

Because this house is located in the middle of the mountain, a simple construction technique was required and we had difficulty selecting the right construction company. Although the construction scale was small, we wanted a competent construction company in the capital rather than a local construction company. Finally, we met a manager of a construction company through the introduction of an acquaintance; we are still really grateful to him even after long the construction work has been completed. Selecting the wrong construction company can result in poor outcomes, even if a design is nicely worked out. Construction lasted eleven weeks. We stayed in lodgings close to the construction site, because of the mountainous location, and brought in the best technicians for each construction phase or process. As a result, the completed details were of a very high quality. For the foundations of the building, we finished the floor using reinforced concrete and applied a light weight wood framing system to the walls and the roof. After that, the roof was finished with zinc and the outer walls were finished with monocouche to neaten up the curved surface of the outer walls. When planning the internal spaces, we decided to create different atmospheres, because the first floor and the second floor have different characteristics. In the shared kitchen we laid porcelain tiles on the first floor, making them easy for the client to clean, while the walls were finished with birch. The kitchen furniture along the curved wall is made of birch so people can gaze at the outside view in front of the sink. When you go up the stairs lined with the birch walls, after taking off your shoes, you reach the spaces on the second floor. This space was finished with diatomite, which is an eco-friendly material. We installed windows in primary positions in all directions so that the occupants can look down the yard or out over the distant mountains.

A Forest Dwelling in a Monumental Landscape

The clearing here in the middle of the mountain is surrounded by the mountain range, and in addition to the spectacular fall foliage, the winter snowy scenes are incredibly beautiful. You can see the surrounding landscape from inside the house and enjoy moments of meditation in nature. Although it is a single volume, the outer walls that coil softly towards the surroundings seem to expand and draw themselves into the forest. On the contrary, the external view, seen through the curved glass inside the house, makes you gaze out into many faces of the forest.

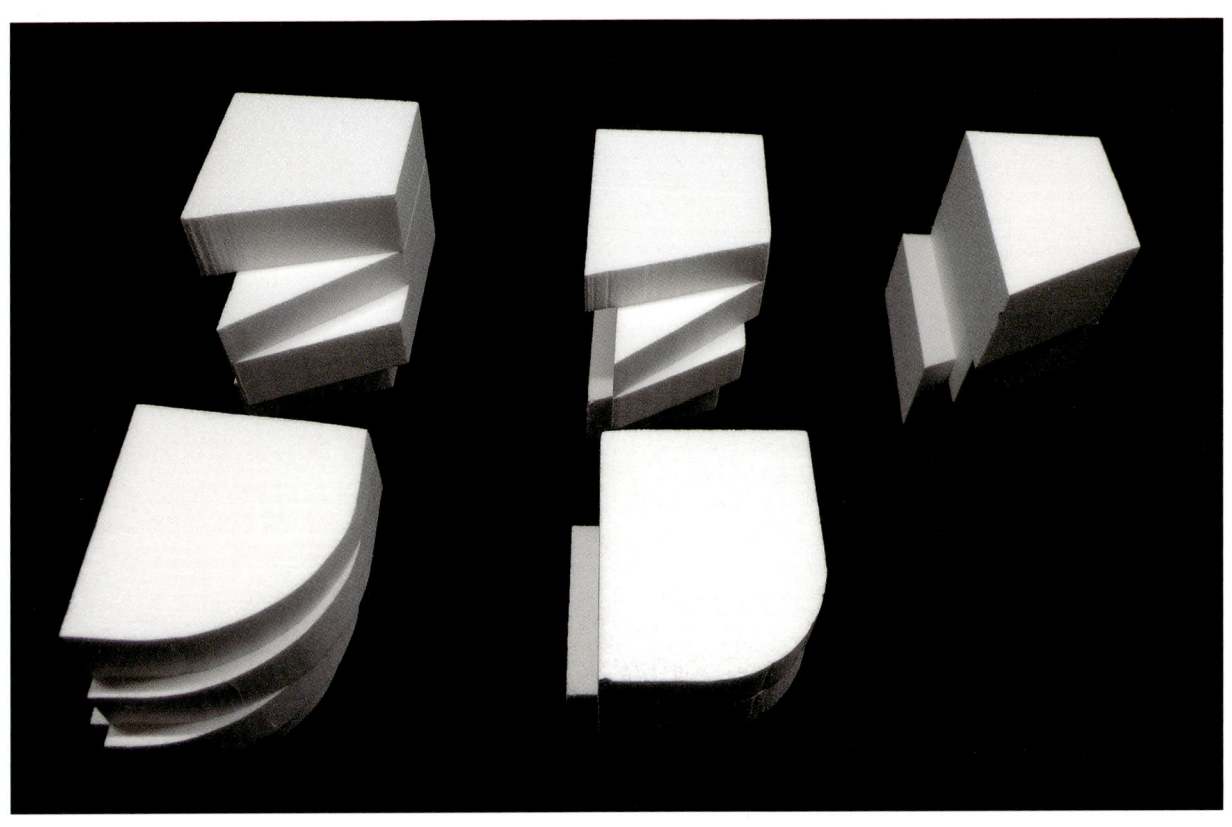

Mass Study

영동주택

한 사람을 위한 작은 집

건축주는 인적이 드문 조용한 마을을 찾아 부모님과 함께 살 집을 지을 땅을 구입하고 설계를 의뢰해왔다. 진입 도로에서 60m 높이의 가파른 경사를 올라오면 산 중턱에 위치한 670㎡ 대지에 80㎡의 농가 주택이 자리 잡고 있었다. 기존 농가 주택은 인테리어를 해서 부모님을 모시고, 본인은 옆에서 혼자 살 집을 짓기로 하고 설계를 시작했다. 새로운 주택의 계획 조건은 1층은 주방 겸 창고로, 2층은 방으로 하되 1층에서 기존 농가 주택과 연결되는 것이었다. 실내이지만 외부공간의 성격을 가지는 1층이 공유공간의 역할을 하고, 2층은 사유공간으로 사용하길 원한 것이다. 사실 처음 이 땅, 그리고 샌드위치 패널 구조의 기존 주택을 보고 어떻게 조화롭게 신축 주택과 배치를 해야 할지, 재료는 어떤 걸 사용해야 할지, 외관은 어떠한 관계성을 가져야 할지 고민이 많았다. '조화가 과연 가능한 걸까?' 결국 기존 주택과는 기능적 동선만 해결하도록 하고 신축 주택은 독자적인 외관을 가지는 것으로 방향을 설정하고 계획을 하였다. 건축주는 미리 평면을 그려 배치와 규모를 가늠하고 각종 친환경 자재와 설비 자료에 대한 연구를 열심히 한 상태였다. 따라서 많은 자료를 살피고 나누며 건축주가 원하는 집에 대한 시공 기술을 함께 고민하며 진행했다. 배치 계획상 건축주가 살 신축 건물은 기존 건물의 우측 안쪽에 자리하게 되었다. 따라서 건축 가능한 면적이 30㎡ 이내인 작은 주택이기에 가구 위치만 바뀌어도 새로운 평면을 구성해야 했고 여러 대안을 발전해가며 건축주와 모형, 3D 모델링 자료를 주고받아 세세하게 도면을 정리했다. 부모님이 사시는 농가 주택의 주방 및 다용도실을 통해 두 주택을 연결하고자 했는데, 대지 경계선과 가까우며 외부 통로까지 고려하다 보니 1층은 20㎡ 정도의 협소한 공간 밖에 없었다. 그래서 2층은 캔틸레버 구조로 1m를 확장하여 공간을 확보하고, 곡면 벽을 활용해 제작 가구를 만들어 실용성을 높였다. 동시에 풍경을 파노라마로 볼 수 있도록 하였다.

열악한 현장과 시공사의 집념

이 주택은 산 중턱에 위치하여 콘크리트 공사는 배제해야 했고, 시공사 선정에도 어려움이 따랐다. 작은 건축이지만 실력 있는 시공사를 찾다가 지인의 소개로 만난 시공 소장은 우리에겐 건물이 완공된 지금에서도 너무나 고마운 사람이다. 설계를 잘해도 시공사를 잘못 만나면 아쉬운 상황이 생기기 마련이다. 공사 기간은 총 11주가 걸렸다. 현장이 산중이기에 인근에 숙소를 정해 놓고 공정별로 훌륭한 기술자들이 시공을 했기에 디테일의 완성도가 아주 높다. 기초는 철근콘크리트로 바닥을 완성하고 벽과 지붕은 경량목구조를 한 후 지붕은 징크, 외벽 마감은 모노쿠쉬를 사용하여 외벽의 곡면을 단정하게 정리하였다. 내부 공간을 계획할 때 1층과 2층의 성격이 다르기에 다른 분위기로 마감하기로 했다. 1층은 공유 주방의 성격을 가져 바닥은 석재 타일을 시공하여 물청소가 가능하도록 하며 벽은 자작나무로 마감하였다. 곡면 벽에 있는 주방 가구는 자작나무로 제작하였으며 싱크대 앞에 서면 바깥 풍경을 볼 수 있다. 신발을 벗고 자작나무 벽을 따라 계단을 오르면 백색의 규조토로 마감한 2층 공간이 펼쳐진다. 동서남북 주요 위치마다 창을 배치하여 마당을 내려다보기도 하고 멀리 산을 볼 수 있다.

숲속에 자리한 작은 건축, 풍경을 담다

건축에서 외관을 만드는 일은 단순히 형태나 파사드를 가리키는 것은 아니고, 내부 공간과 외부 공간의 경계가 어떻게 구성되어 공간을 구축하고 있는지를 정의하는 것이다. 외부 공간과 내부 공간이 어떤 식으로 공간의 경계를 구성하고 있는지는 내부와 외부의 관계성에서 비롯된다. 이 주택의 외관은 땅의 경계로부터 빚어지고 내부 공간과 외부 공간은 곡면의 창호를 통해 관계를 맺고 있다. 대지는 산으로 둘러싸여 있으며 가을 단풍, 특히 설경이 매우 아름다운 곳이다. 주변의 풍경을 주택 내부로 끌어들이고 자연에 둘러싸인 채 명상을 즐길 수 있는 집이다. 단일한 덩어리의 집이지만 주변을 향해 부드럽게 감아 도는 외벽은 숲으로의 확장과 연계를 유연하게 받아들이는 모습이고, 반대로 내부에서 곡면 유리를 통해 바라보는 외부의 전경은 다양한 숲의 표정에 시선을 던지게 한다.

1. ZINC FINISH	1. 징크마감
2. T2 SHEET WATERPROOF	2. T2 쉬트방수
3. OSB 4'×8' 11.1MM	3. OSB 4'×8' 11.1MM
4. RAFTER VENT	4. RAFTER VENT
5. R30 'C' GRADE T220 INSULATION	5. R30 다등급 T220 인슐레이션
6. S.P.F 2×8 @600	6. S.P.F 2×8 @600
7. T9 STEEL PLATE FLASHING	7. T9 스틸플레이트 후레싱
8. MONOCOUCHE(EXTERNAL INSULATION SYSTEM)	8. 모노쿠쉬(외단열시스템)
9. T9 STEEL PLATE / URETHANE PAINT	9. T9 스틸플레이트 / 우레탄 페인트
10. T12 BIRCH	10. T12 자작나무
11. T12 GYPSUM BOARD	11. T12 석고보드
12. PVC SYSTEM WINDOW	12. PVC 시스템창호
13. APP' HARDENED FLOORING	13. 지정 강화마루
14. T75 PANEL HEATING	14. T75 판넬히팅
15. S.P.F 2"×6" @400	15. S.P.F 2"×6" @400
16. R21 'B' GRADE T140 INSULATION	16. R21 나등급 T140 인슐레이션
17. T11.1 OSB PLYWOOD	17. T11.1 OSB 합판
18. MOISTURE-PERMEABLE AND WATERPROOF	18. 투습방수지(타이백)
19. MONOCOUCHE	19. 모노쿠쉬
20. CONCRETE POLISHING	20. 콘크리트 폴리싱
21. AL INSULATED WINDOWS	21. AL 단열창호

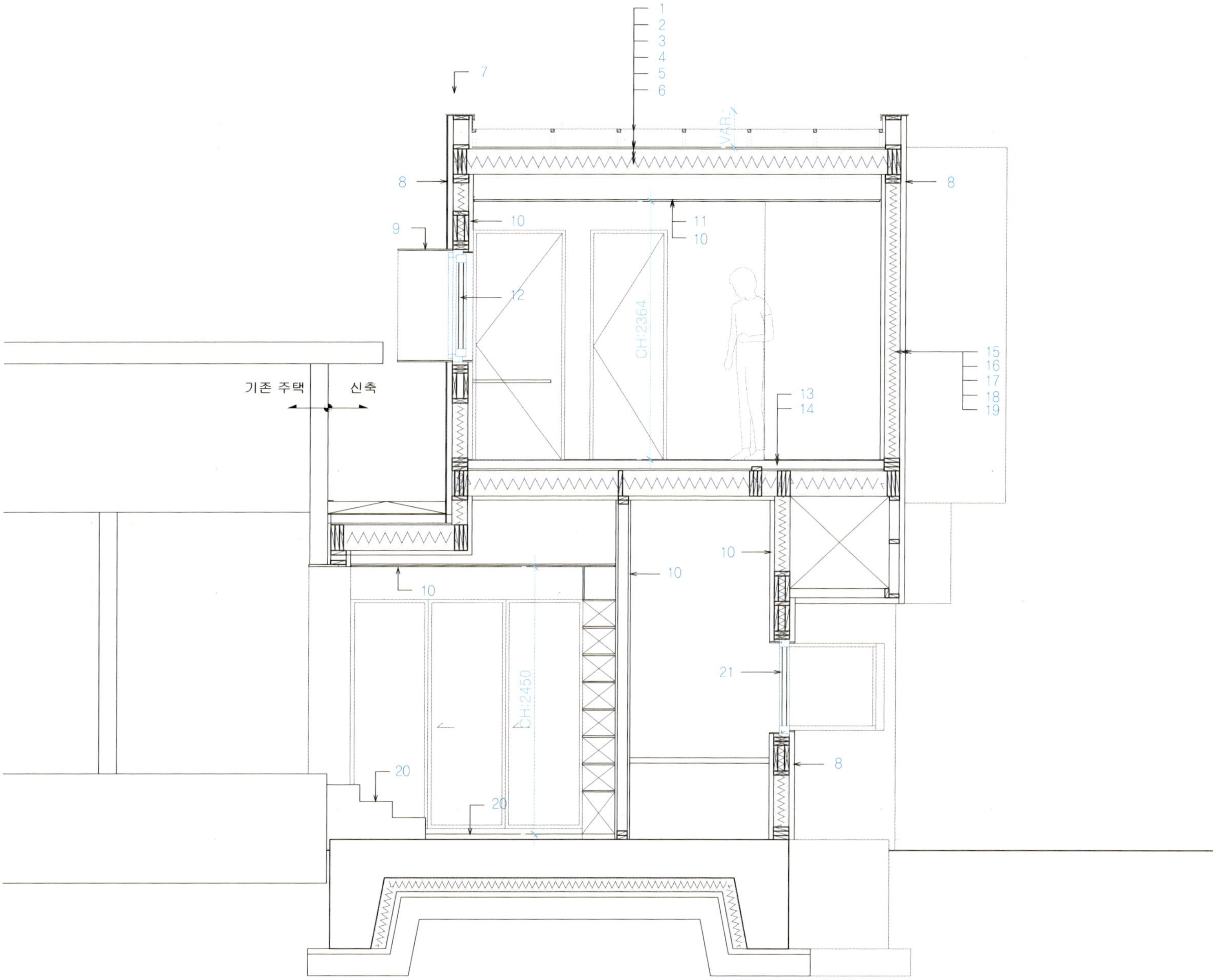

Detail

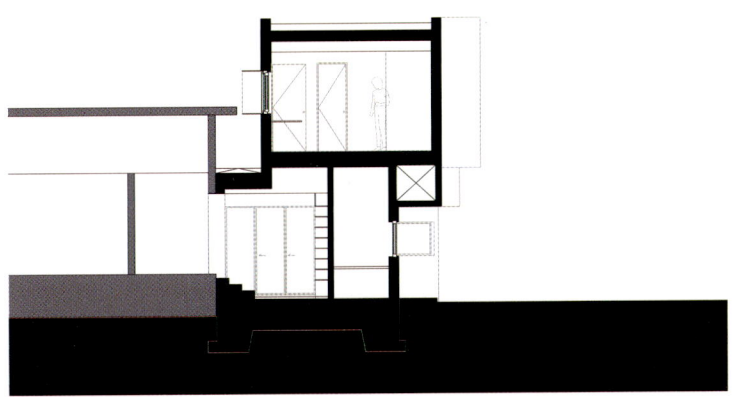
Section A

Section B

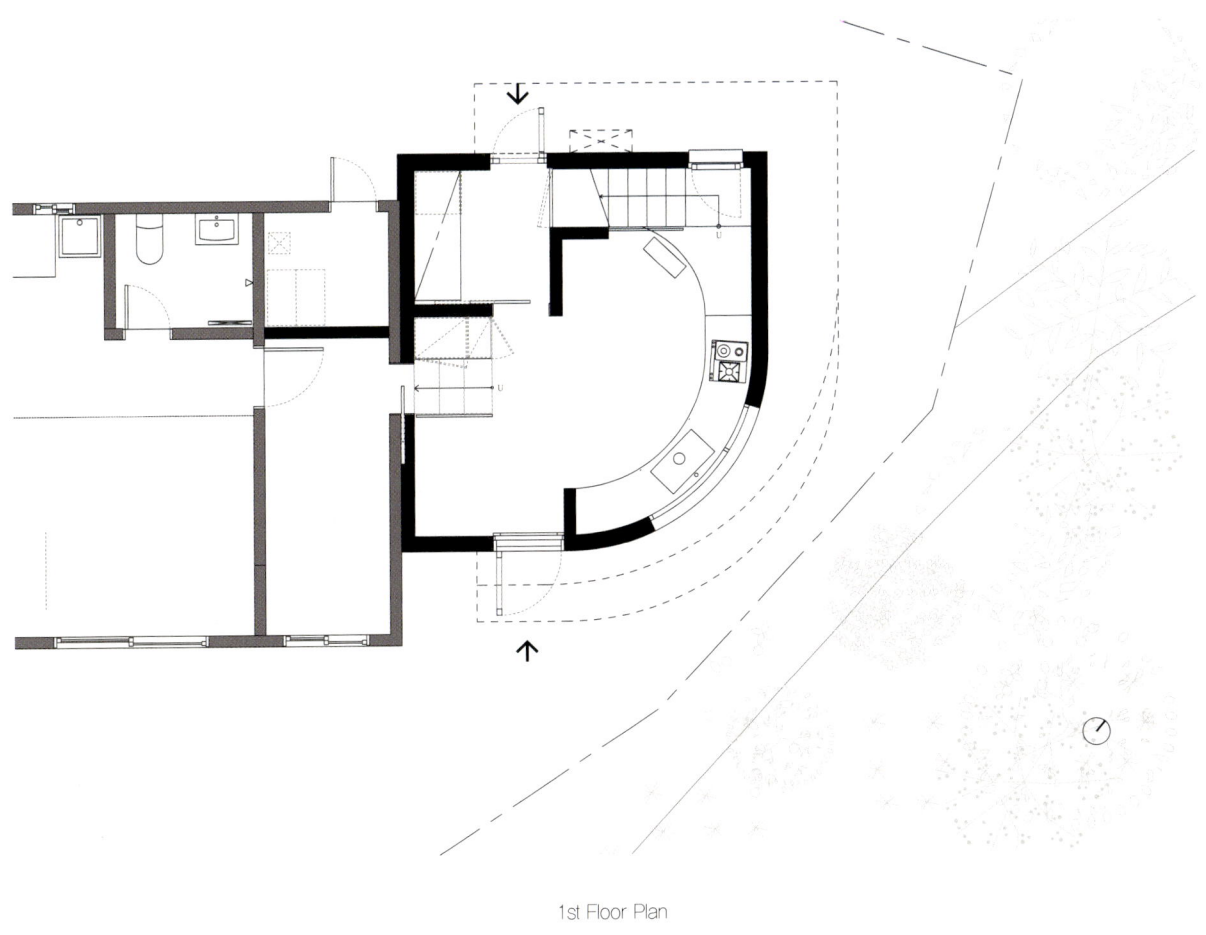

1st Floor Plan

Maengdong House Eumseong, Korea

How Architects Inc.

Architects How Architects Inc. **Project Team** Sebeom Oh, Namin Hwang **Location** Eumseong, Korea
Site Area 329㎡ **Bldg. Area** 52.83㎡ **Bldg. Scale(Floor)** 2FL **Structure** Reinforced Concrete **Max. Height** 7m
Exterior Finish White Old Brick **Construction** MOA **Photographer** Taeksoo Lee

설계 ㈜하우건축사사무소 **설계참여자** 오세범, 황남인 **위치** 충북 음성군 맹동면 마산리 **대지면적** 329㎡ **건축면적** 52.83㎡ **규모** 지상2층 **구조** 철근콘크리트 **최대높이** 7m **외부마감** 백고벽돌 **시공** 엠오에이 **사진** 이택수

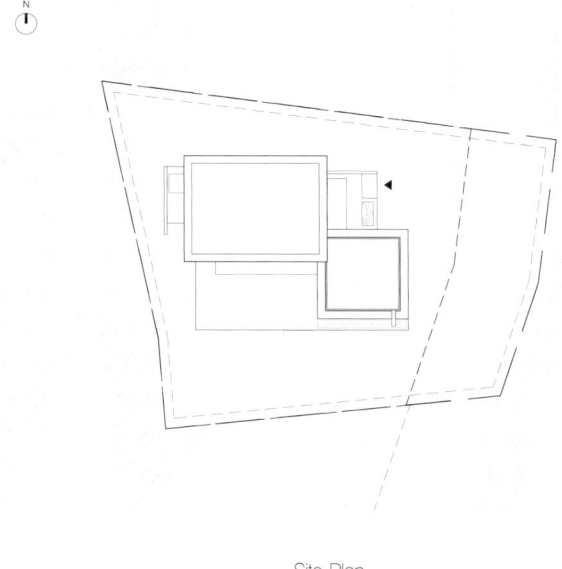

Site Plan

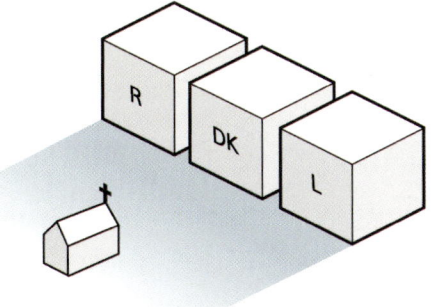 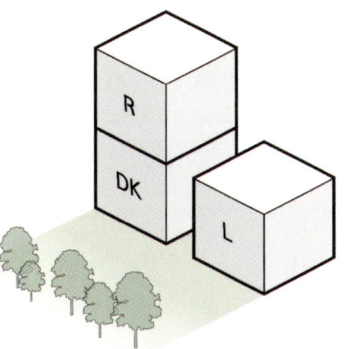 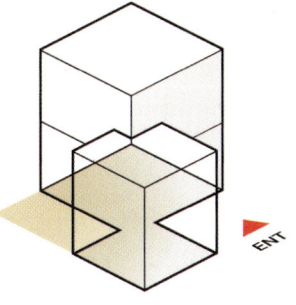

Concept Diagram

Maengdong House is a weekend house located in a rural village in North Chungbook Province. On the western side of the site, there is a long vegetable garden that the owner couple puts all their heart into every weekend, and there is a big road outside. To the east of the site, large and small fields of the village are lined up with the backdrop of a distant mountain range.

About 200m to the south, there is a small hill, and on the top of the hill, a straight red brick church building is standing facing this house. A bell tower rises in the center of the cathedral, and a small statue of Our Lady in front of it overlooks the yard of the house. Coincidentally, the client was a strong Catholic. Initially, the house was designed to face the cathedral, but later the direction of the house was slightly changed at the request of the owner. It was because he felt it was a little disrespectful to be facing Our Lady who was looking down.

The client wanted a frugal house with minimal space. A living room that doubles as a guest room, a small kitchen and dining room, and a second floor room with bathroom. In order to make the exterior of the building as simple as possible, each space is planned as a single module. When the second-floor room was raised above the dining room, two cubes, high and low, were formed. The two masses were placed facing south, but they were not placed side by side, but were placed opposite each other. Two small external spaces were created between the two displaced masses. One became the entrance space in front of the entrance to the house from the north road, and the other became the courtyard connecting the living room and dining room. The windows were all planned with a clear purpose. . The living room window was raised so that the interior could not be seen from the road and was torn horizontally so that the eastern ridge could be seen as a panoramic view. In the stairwell, the low sunlight in the late afternoon fell on the wall through the high side window, and the window in the bathroom was used as a frame to capture the scenery.

The highlight of this house is by far the landscaping. When planning the external space, only the area was planted, and the owner selected the tree species and planted them in the construction stage. In particular, in the large yard on the south side of the house, the garden was completed with trees and flowers planted in advance by the owner several years ago.

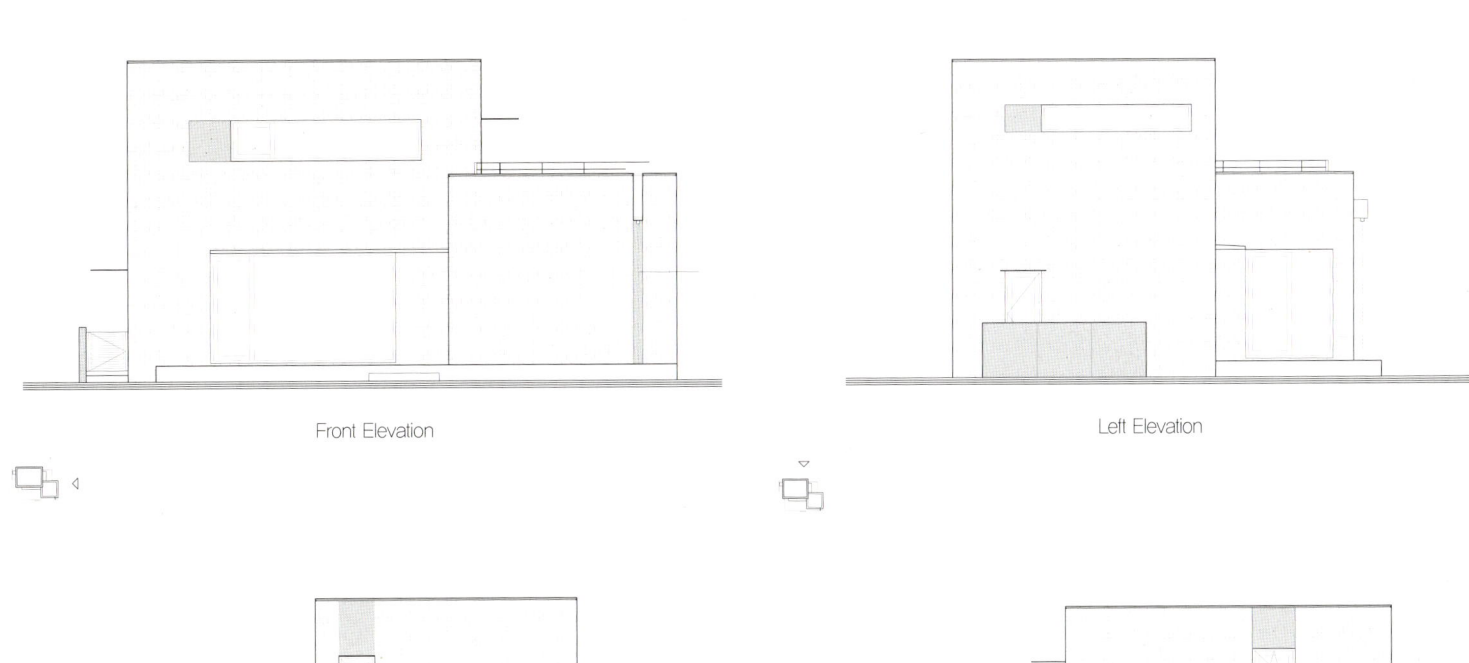

Front Elevation

Left Elevation

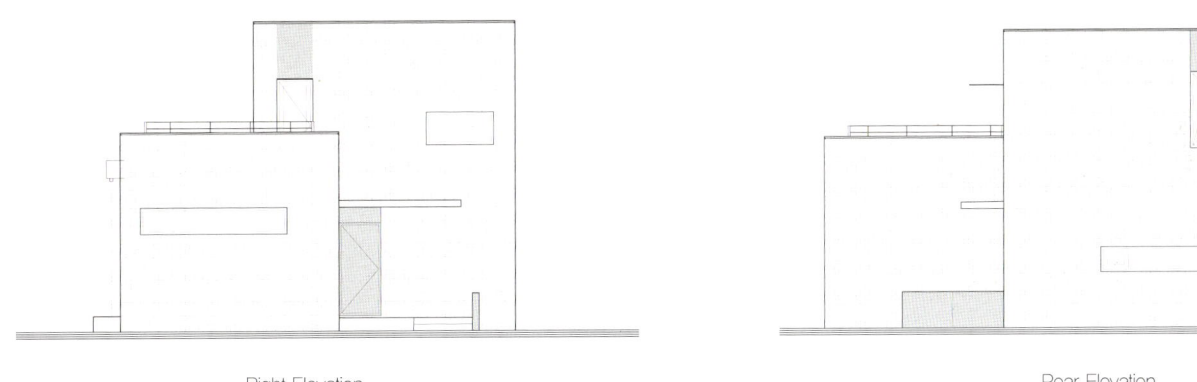

Right Elevation

Rear Elevation

맹동주택

맹동주택은 충북 한 시골마을에 자리한 주말주택이다. 대지의 서쪽 편에는 건축주 부부가 주말마다 정성을 쏟는 긴 텃밭이 있고 그 바깥쪽으로는 큰 도로가 있는데, 텃밭과 도로 사이에는 길이 나면서 잘리고 남은 좁은 숲이 있어 외부의 시선에서 대지를 보호해준다. 대지 동쪽으로는 멀리 펼쳐진 산자락을 배경으로 마을의 크고 작은 밭들이 웅긋중긋하게 늘어서 있다. 남쪽으로 200m 남짓 떨어진 곳에는 조그만 언덕이 있는데, 언덕 꼭대기에 붉은 벽돌로 반듯하게 지어진 성당 건물이 이 집을 바라보고 서있다. 성당의 중앙에는 종탑이 솟아 있고 그 앞에 세워진 작은 성모상이 집의 마당을 내려다보고 있다. 공교롭게도 건축주는 견실한 천주교 신자였다. 당초 집이 성당을 곧바로 바라보도록 설계하였는데, 후에 건축주 요청으로 집의 방향을 약간 틀었다. 내려다보는 성모님을 정면으로 마주보고 있는 것이 조금 불경하다고 느낀 탓이었다.

건축주는 최소한의 공간으로 구성된 검박한 집을 원했다. 게스트룸을 겸한 거실, 작은 주방과 식당, 욕실이 딸린 2층 방 하나. 건물의 외관을 최대한 단순하게 만들기 위해 각 공간을 하나의 모듈로 계획했다.

2층 방을 식당 위에 올리니, 높고 낮은 두 개의 큐브가 형성됐다. 남쪽을 향해 두 매스를 배치하되, 나란히 놓지 않고 서로 어긋나게 두었다. 어긋난 두 매스 사이에 두 개의 작은 외부공간이 생겼는데, 하나는 북쪽 길에서 집으로 들어오는 현관 앞 진입공간이 되었고 나머지 하나는 거실과 식당을 연결하는 안마당이 되었다.

창들은 모두 명확한 목적을 가지고 계획되었다. 거실 창은 길에서 내부가 들여다보이지 않게 높이면서 가로로 길게 찢어 동쪽 산등성이를 파노라마처럼 볼 수 있게 하였고, 2층 방의 창은 남쪽의 채광을 확보하면서 성당을 바라볼 수 있게 하였다. 계단실에는 고측창을 통해 늦은 오후의 낮은 햇빛이 벽면에 떨어지게 하였고, 욕실의 창은 풍경을 담는 액자처럼 활용하였다.

이 집의 백미는 단연코 조경이다. 외부공간을 계획하면서 식재는 그 영역만 잡아놓았는데, 시공단계에서 건축주가 직접 수종을 골라 식재하였다. 특히 집 남쪽의 넓은 마당에는 건축주가 몇 년 전부터 미리 심어놓은 나무들과 풀꽃들이 어우러져 정원이 완성되었는데, 멀리 보이는 성당의 모습과 포개지며 아름다운 풍경을 만들어낸다.

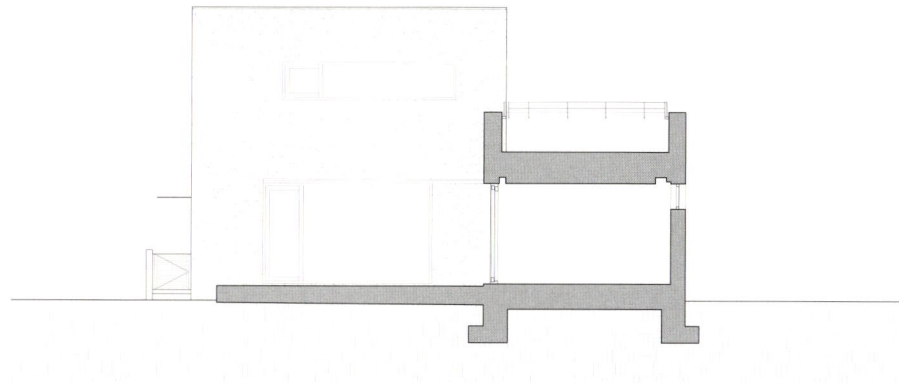

Section I

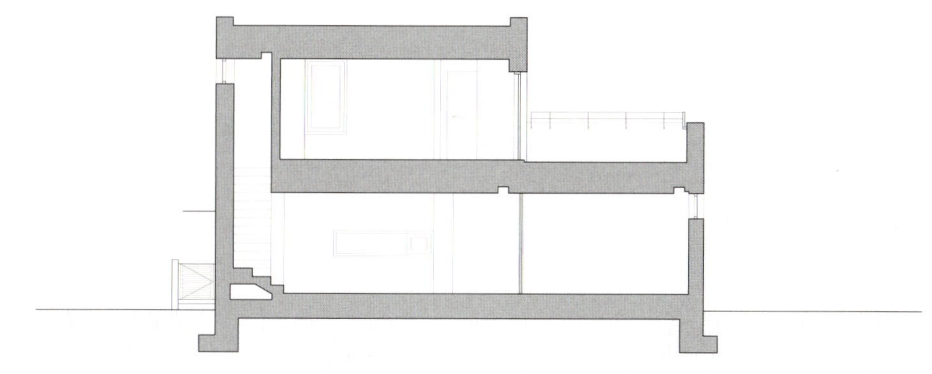

Section II

1. THK130 PF보드
2. ㅁ-30×30 목재천장틀
3. T9.5 석고보드2겹
4. 전면퍼티 후 지정 수성페인트
5. THK10 강마루
6. THK50 시멘트몰탈(온수온돌)
7. THK40 경량기포콘크리트
8. THK30 비드법보온판 가호
9. 침투성방수 위 보호몰탈
10. 노출콘크리트 면정리
11. 발수제 도포

1. THK130 PF BOARD
2. ㅁ-30×30 WOODEN CEILING FRAME
3. 2 PLY OF T9.5 GYPSUM BOARD
4. APP' WATER-BASED PAINT AFTER FRONT PUTTY
5. THK10 KANG MARU
6. THK50 CEMENT MORTAR(HOT WATER ONDOL)
7. THK40 LIGHTWEIGHT AERATED CONCRETE
8. THK30 BEAD METHOD THERMAL INSULATION PLATE
9. PERMEABLE WATERPROOF ON PROTECTIVE MORTAR
10. EXPOSED CONCRETE FACE
11. WATER REPELLENT APPLICATION

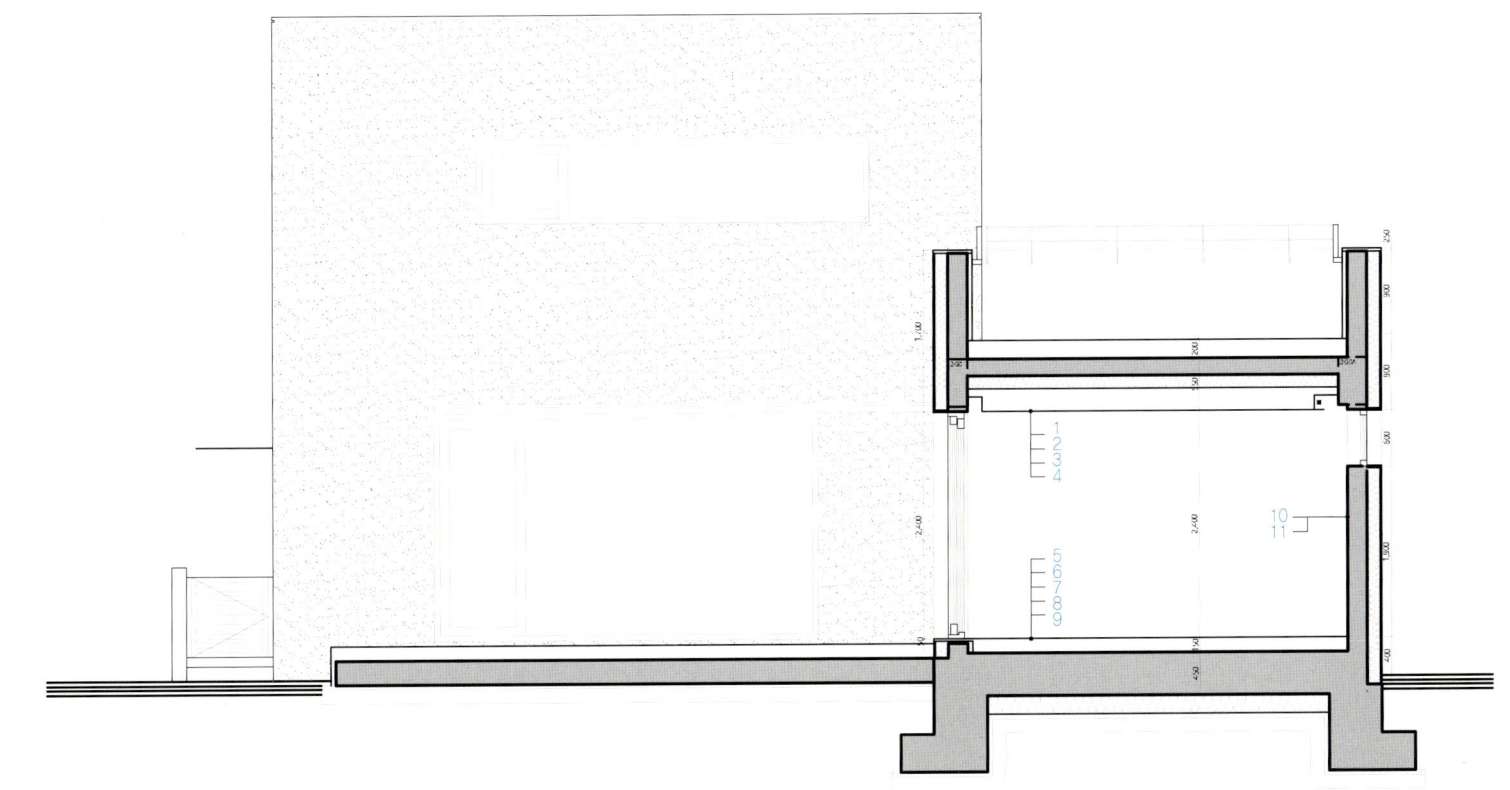

Section Detail

1. ENTRANCE
2. LIVING ROOM
3. KITCHEN
4. BATHROOM
5. STAIRS
6. UTILITY ROOM
7. BED ROOM
8. TERRACE

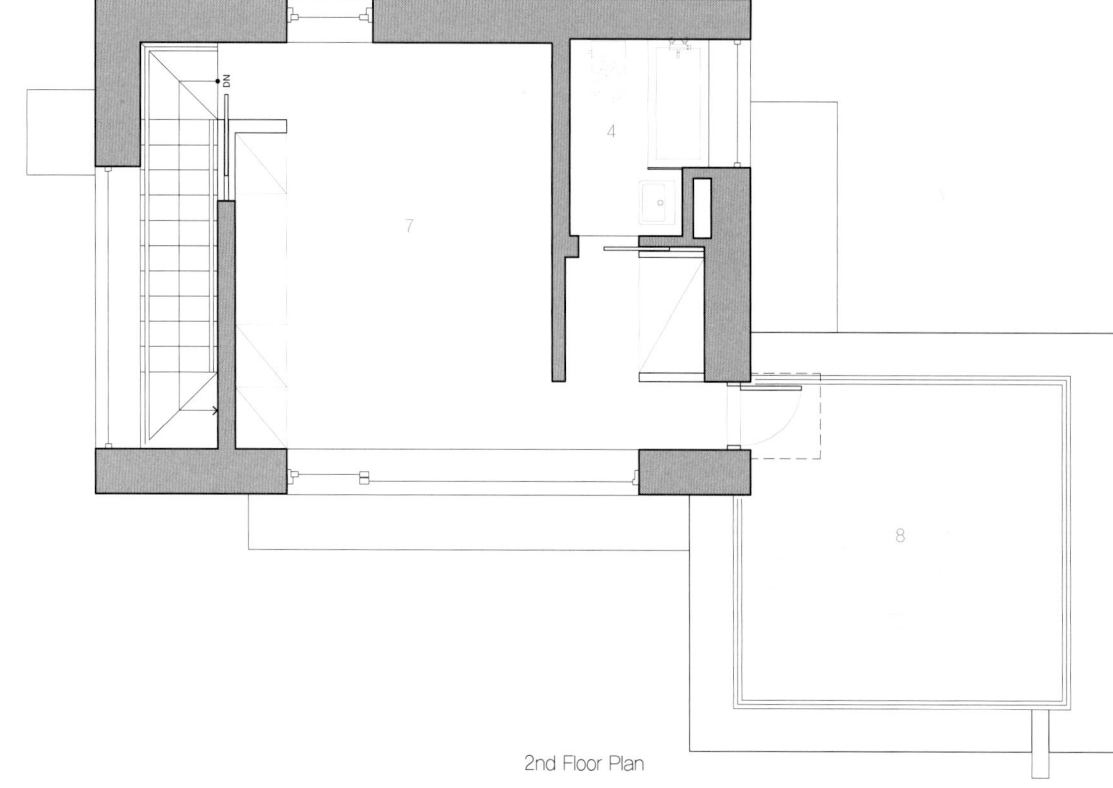

2nd Floor Plan

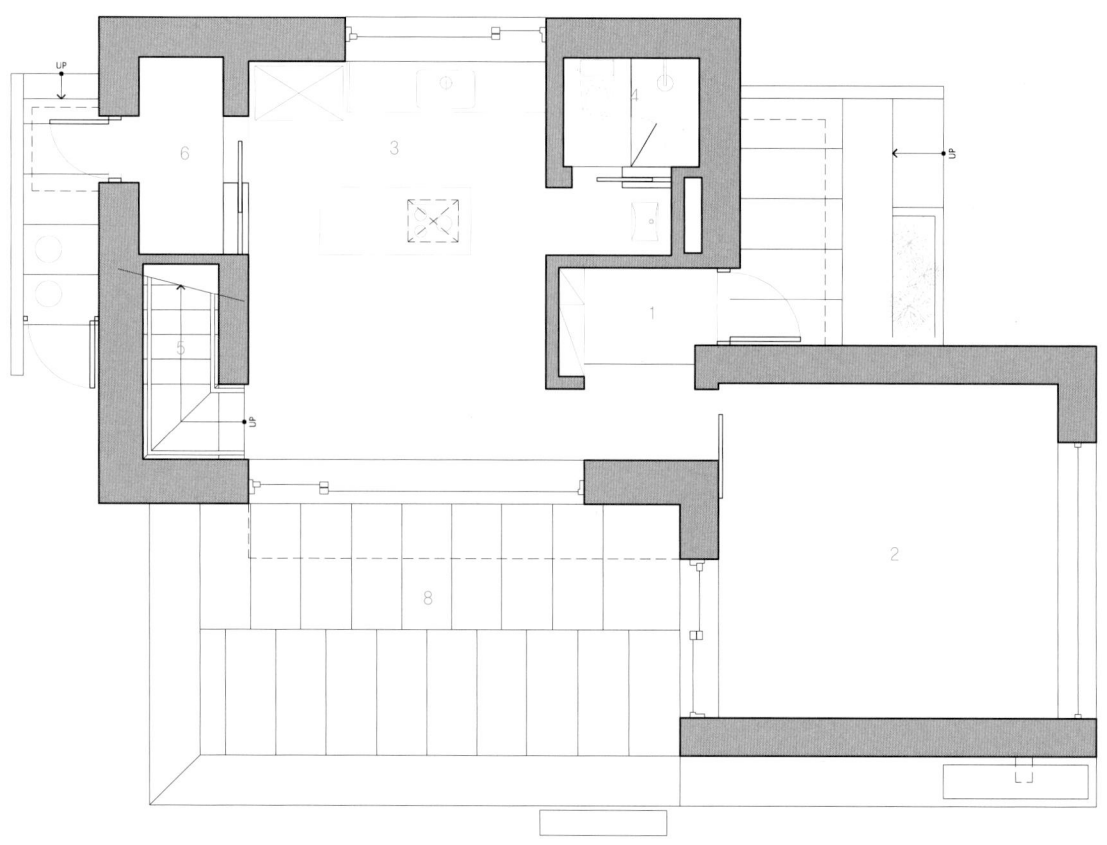

1st Floor Plan

25

Landschaft Yeoju, Korea

Architects601

Architects Architects601 / Keunyoung Shim **Location** Yeoju, Korea **Site Area** 578㎡ **Bldg. Area** 98.99㎡ **Gross Floor Area** 138.63㎡ **Bldg. Scale(Floor)** 2FL **Structure** Reinforced Concrete **Max. Height** 7.1m **Exterior Finish** Stucco Flex(EPS 50mm) **Photos Offer** Architects601

설계 아키텍츠601 / 심근영 **위치** 경기도 여주시 걸은리 **대지면적** 578㎡ **건축면적** 98.99㎡ **연면적** 138.63㎡ **규모** 지상2층 **구조** 철근콘크리트 **최고높이** 7.1m **외부마감** 스타코플렉스 마감(EPS 50mm) **사진제공** 아키텍츠601

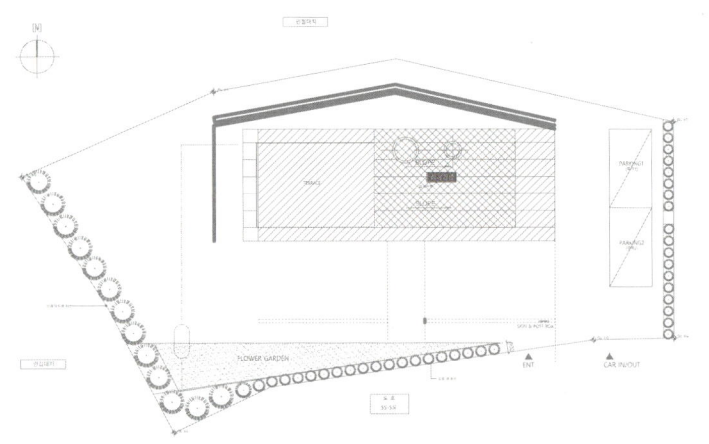

Site Plan

Elevation

This is a detached house designed for elderly parents who are away from their children in a provincial city. The owner came to us in hopes that the elderly parents would spend their old age comfortably and comfortably in a natural site with fresh air and beautiful scenery. He wanted a level of solid and solid design and finishing materials that was easy to manage even after 10 or 20 years and had no defects in durability. wanted to be And since we wanted to bring another joy of life to our parents, who have lived in the apartment for a long time, through the house we built ourselves, we wanted to realize the shape of a white and bright house that is like the joy of the heart. It reminded me of Le Corbusier's Villa Savoye house. On the other side of the road, the old house that stood on the road is reborn as a house of joy, embraced by the earth in the shape of a white house.

The geometrical architectural form that contrasts with nature was enough to become a pure white drawing paper that embraces nature with a clean off-white Starcoplex finish. On the white drawing paper, the deep honesty of the wood grain and the deep shadows of the shadows cast by the sun reflect the changes of the four seasons, and the landscape as colorful as the passage of time is embraced by the owner's family. This house was born as a 'House of Joy', hoping that it will become a comfortable and comfortable home in life with the family's time and wishes.

The concept of this house is a representation of minimalistic 'modernism' and contains the language of the classic concept 'axis and symmetry' at the same time. It was enough to capture both function and aesthetics at the same time as a homage to the 'white house' that the client wanted and the Villa Savoye house, which came to mind intuitively after seeing the site as a designer. Minimalism is expressed through the proportion and repetition of the window opening, and the play of shadows according to the shade of light honestly reveals the flow of nature in the monotony of architecture, adding to the depth of white architecture.

The main entrance and exit of the axis supporting the center match the landscape of the mountain behind, bringing down the light of the sky and penetrating the panorama of light into the space. Residents experience the encounter of light and space, the purification of air, and the diversity of space in the course of various passages of time.

1. APP' ZINC PANEL FIN.
2. SHEET WATERPROOF INSTALLATION
3. THK12 WATERPROOF PLYWOOD
4. SQUARE PIPE 20×40 @450
5. THK180 EXTRUDED POLYSTYRENE SHEET
6. WOOD STRUCTURE CEILING FRAME INSTALLATION
7. THK9.5 GYPSUM BOARD 2PLY
8. APP' WALLPAPER FIN.
9. APP' ELECTRIC ONDOL FLOOR FINISH
10. T30 INSULATION
11. APP' ZINC PANEL(GRAY)
12. APP' STUCCO, MONOCOAT FIN.
13. T50 SOUND ISOLATION / INSULATION
14. APP' TILE FINISH
15. APP' ONDOL FLOOR FLOORING FINISH
16. BOILER INSTALLATION(INCLUDING INSULATION T30)

1. 지정 징크판넬마감
2. 시트방수 취부
3. THK12 내수합판
4. 각파이프 20×40 @450 취부
5. THK180 압출법보온판
6. 목구조 천장틀 취부
7. THK9.5 석고보드 2겹 마감
8. 지정 도배마감
9. 지정 전기온돌마루 마감
10. 단열재 T30 포함
11. 지정 징크판넬(그레이)
12. 지정 스타코, 모노코트마감
13. T50 차음/ 단열재
14. 지정 타일마감
15. 지정 온돌마루 플로링 마감
16. 보일러 취부(단열재 T30 포함)

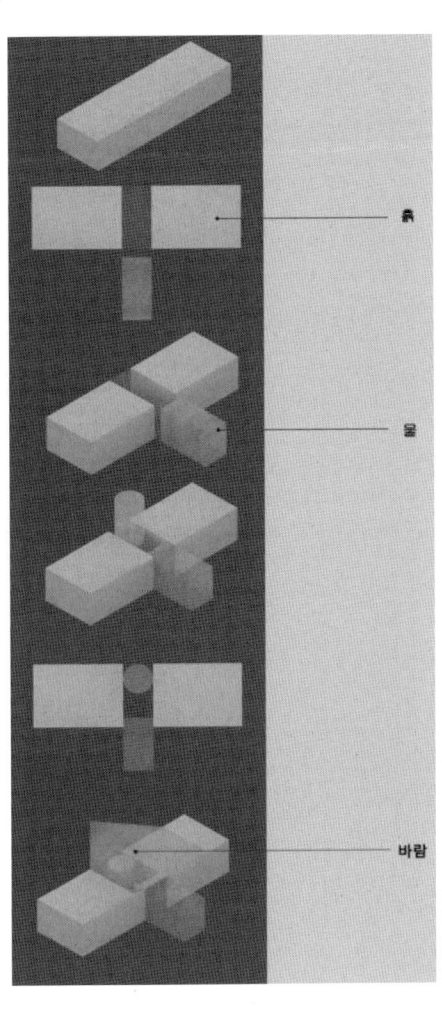

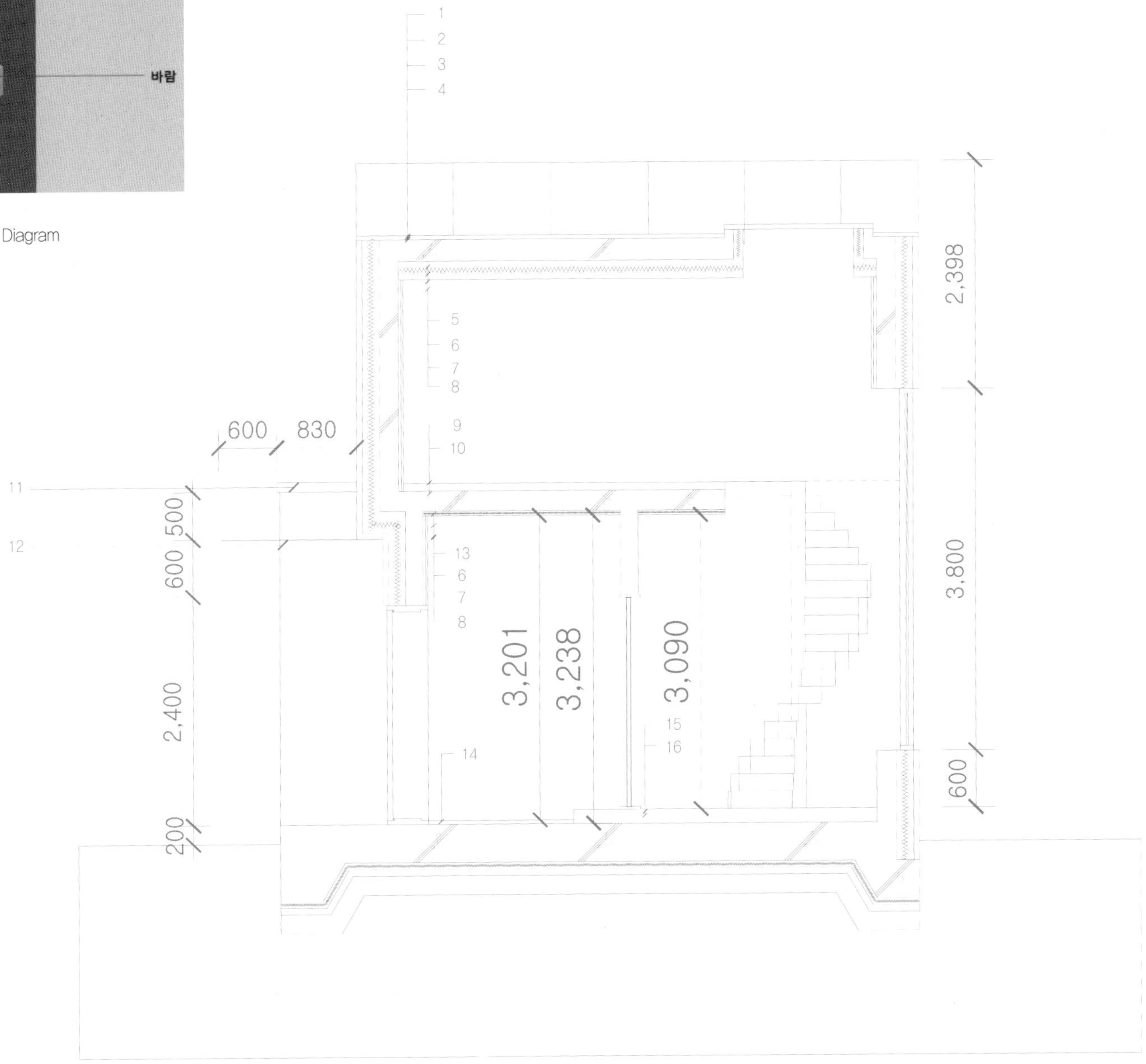

Concept Diagram

1. APP' ZINC PANEL FIN.
2. SHEET WATERPROOF INSTALLATION
3. THK12 WATERPROOF PLYWOOD
4. SQUARE PIPE 20×40 @450
5. THK180 EXTRUDED POLYSTYRENE SHEET
6. WOOD STRUCTURE CEILING FRAME INSTALLATION
7. THK9.5 GYPSUM BOARD 2PLY
8. APP' WALLPAPER FIN.
9. APP' ELECTRIC ONDOL FLOOR FINISH
10. T30 INSULATION
11. APP' ZINC PANEL(GRAY)

1. 지정 징크판넬마감
2. 시트방수 취부
3. THK12 내수합판
4. 각파이프 20×40 @450 취부
5. THK180 압출법보온판
6. 목구조 천장틀 취부
7. THK9.5 석고보드 2겹 마감
8. 지정 도배마감
9. 지정 전기온돌마루 마감
10. 단열재 T30 포함
11. 지정 징크판넬(그레이)

Section Detail

여주 기쁨의 주택

지방 도시에서 자녀들과 떨어져 지내고 계신 노부모님을 위해 계획된 단독주택으로 건축주는 노부모님이 공기 좋고 풍광이 아름다운 자연의 터에서 노후를 편안하고 안락하게 보내시길 바라며 우리들을 찾아왔다. 10년, 20년이 지나도 관리가 용이하며, 내구성에 하자가 없고 탄탄하고 견고한 설계와 마감재의 수준을 원하셨고, 단층의 주택보다는 다락방을 2층으로 두어 공간의 다양한 쓰임과 풍경들을 마주할 수 있게 되기를 원하였다. 그리고 오랫동안 아파트에 거주하셨던 부모님께 직접 손수 지은 주택을 통해 삶의 또다른 기쁨을 안겨드리고 싶었기에 마음의 기쁨과도 같은 하얗고 화사한 집의 형상을 구현해 주길 원하였고, 우리는 건축주의 바램과 이야기를 들으며 르꼬르뷔제의 빌라 사보아 주택을 연상하게 되었다. 그렇게 도로 이면 길 위에 서 있던 오랜 구옥은 하얀 집의 형상으로 대지의 품을 안으며 기쁨의 주택으로 새롭게 태어난다.

자연과 대비되는 기하학적 건축 형태는 깨끗한 미색의 스타코플렉스 마감으로 자연을 품어내는 순수무구한 백색의 도화지가 되기에 충분했다. 그 백색 도화지에 나무결의 깊은 우직함과 태양빛에 드리워진 그림자의 짙은 음영으로 사계절의 변화를 담고 시간의 흐름만큼 다채로운 풍경으로 건축주 가족의 품에 안긴다. 그렇게 가족들의 시간과 바램을 담아 생활 속에 안락하고 편안한 보금자리가 되어주길 바라며 본 주택은 '기쁨의 주택'으로 태어났다.

이 주택의 컨셉은 미니멀한 '모더니즘'을 표상으로 고전적인 개념인 '축과 대칭'에 대한 언어를 동시에 담고 있다. 그것은 건축주가 바랬던 '하얀 집'과 설계자로서 대지를 보고 직관적으로 떠오른 빌라 사보아주택에 대한 오마주로서 기능과 미학을 동시에 담아내기에 충분했다. 창의 개구부의 비례와 반복을 통해 미니멀리즘을 표현하며, 빛의 음영에 따른 그림자의 유희는 건축의 단조로움에 자연의 흐름을 정직하게 드러내며 백색 건축의 깊이를 더해 준다.

중심을 지탱하는 축의 주출입구는 내부에서 뒷산의 풍경과 일치되어 하늘의 빛을 끌어내리고 빛의 파노라마를 공간으로 관입시킨다. 거주자는 다양한 시간의 흐름 가운데 빛과 공간의 조우, 공기의 순화를 경험하며, 공간의 다양성을 마주한다.

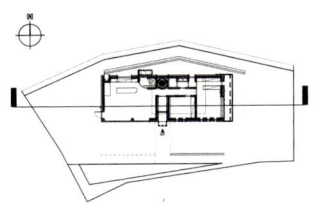

1. KITCHEN & DINING ROOM
2. LIVING ROOM
3. PANTRY ROOM
4. GUEST ROOM
5. MASTER ROOM
6. TERRACE
7. GARRET ROOM

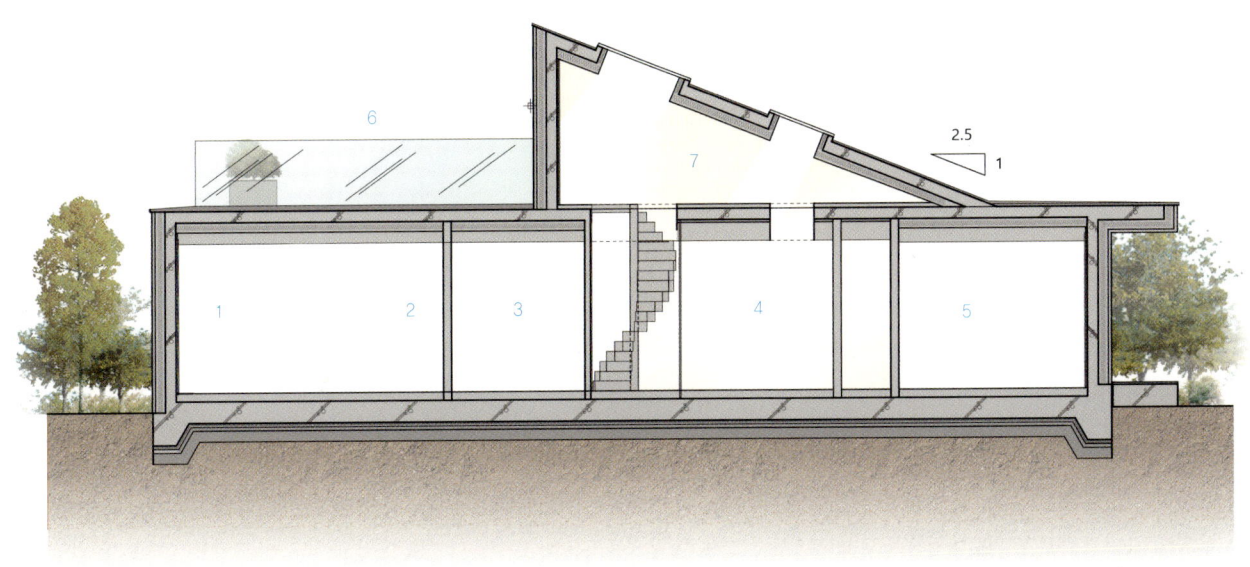

Section

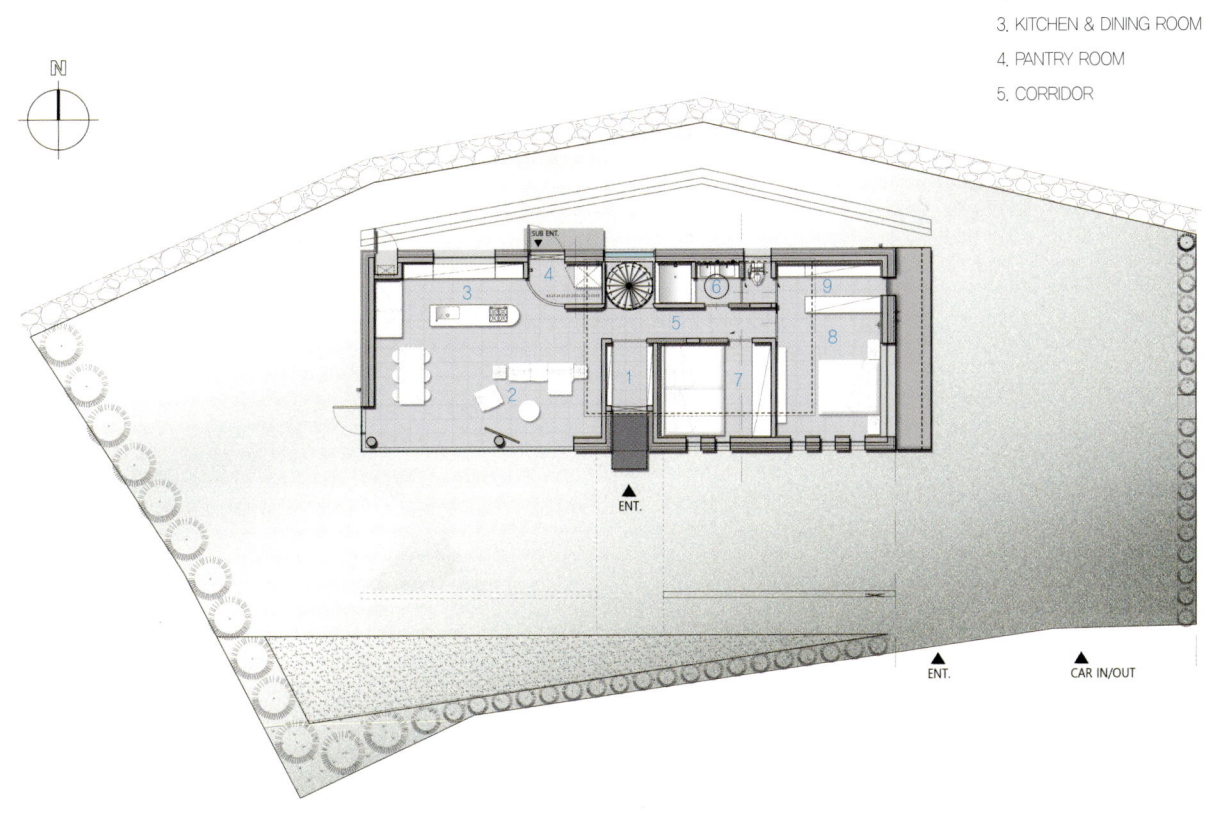

1. ENTRANCE
2. LIVING ROOM
3. KITCHEN & DINING ROOM
4. PANTRY ROOM
5. CORRIDOR
6. BATH ROOM
7. GUEST ROOM
8. MASTER ROOM
9. DRESS ROOM

1st Floor Plan

1. TERRACE
2. GARRET ROOM
3. STORAGE

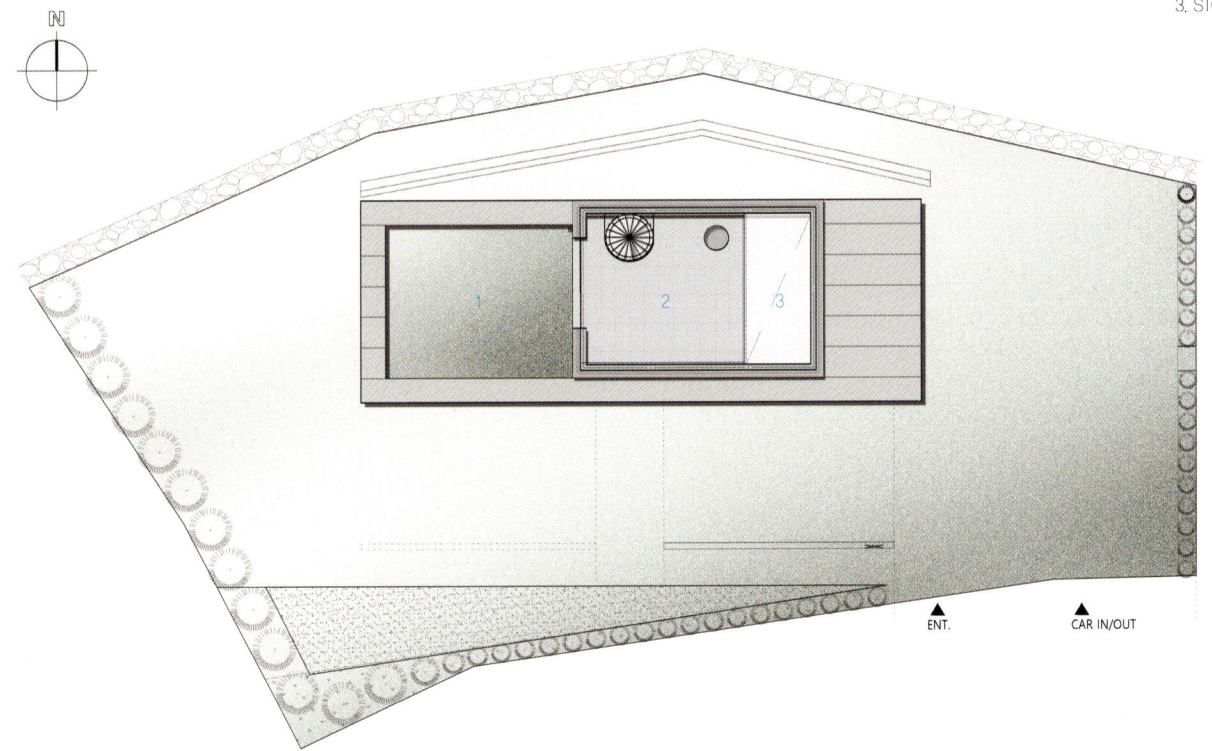

2nd Floor Plan

ENT.   CAR IN/OUT

Ger House Ganghwa, Korea

atelierjun

Architects atelierjun / Junsang You **Project Team** Junhyung Lee **Location** Ganghwa, Korea **Site Area** 774㎡ **Bldg. Area** 99.59㎡ **Gross Floor Area** 99.76㎡ **Bldg. Scale(Floor)** 2FL **Structure** Reinforced Concrete **Max. Height** 7.1m **Exterior Finish** Stucco-Flex, Red Cedar **Client** Junryeol Park **Photos Offer** atelierjun

설계 아뜰리에쥰 건축사사무소 / 유준상 **설계참여자** 이준형 **위치** 인천광역시 강화군 길상면 길직리 1268-1 **대지면적** 774㎡ **건축면적** 99.59㎡ **연면적** 99.76㎡ **규모** 지상2층 **구조** 철근콘크리트 **최대높이** 7.1m **외부마감** 스타코플렉스, 적삼목 **발주처** 박준렬 **사진** 아뜰리에쥰 건축사사무소

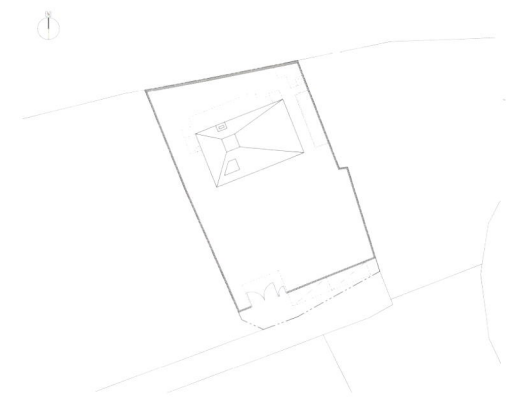
Site Plan

Located in Giljik-ri, Ganghwa Island, Ger House is a weekend house for client who does business in Incheon. The house site is an area where fields are cleared and developed as residential areas, and is located at the highest point on gentle slopes. The client wanted to build it as small as a national life house of less than 100 square meters. All rooms and living rooms are on the south side, except for the red clay Korean dry sauna that enters from the outside. Each room is placed on the left and right sides of the living room and kitchen. At the beginning of the plan, the architect wanted to set the first floor at 100 and the second floor at an attic not included in the total floor area to make up for the insufficient area, but due to regulations, he changed his plan to a small two-story house instead of an attic. In these small houses, the shape of the roof was a very important factor. The roof of Ger house is not a roof that is tilted in a particular direction like a simple garble roof. The four-sided walls are tilted to the center to emphasize the living room, which is the center of the space, and a skylight is installed there to make the central space even brighter through natural lighting. The living room, which overlooks the south garden, where the client has carefully cultivated, creates a cozy atmosphere with a high level of openness and light coming through the skylight. The interior is finished in white, making the shade of light that changes over time stand out.

On the southern side of the tree, the wooden floor of the traditional Korea house (toenmaru) is located in the shape of a house. Roofs without eaves in the form of buildings increase the risk of leakage at the top of windows, so large windows in the south are located inside the wooden floor (toenmaru). The red clay room followed the classical the classical way of heating the room by removing firewood from the furnace, and wood-patterned tiles were used. The chimney has a concise form along the shape of the building, and serves as a point by dropping shadows on the back of the plain building

After the structure of the building was erected, people in the neighborhood called it 'Mongolian tents'. The ger, a traditional Mongolian residential form, is a space where people where people live around an opening in the center, and there are many similarities to the structure of this house. The architect named it the ger house. The client said that the apartment with a much larger floor space was rather frustrating after completion, and that he almost lived there.

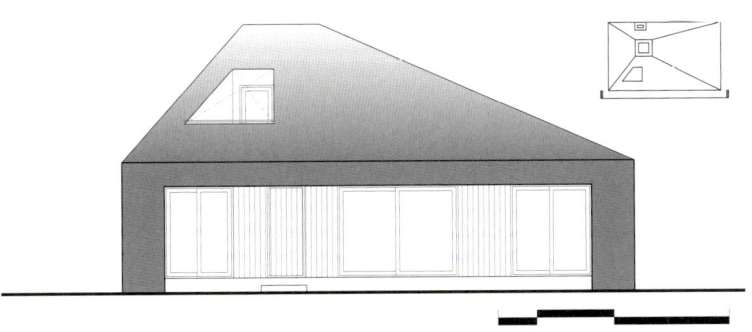

South Elevation

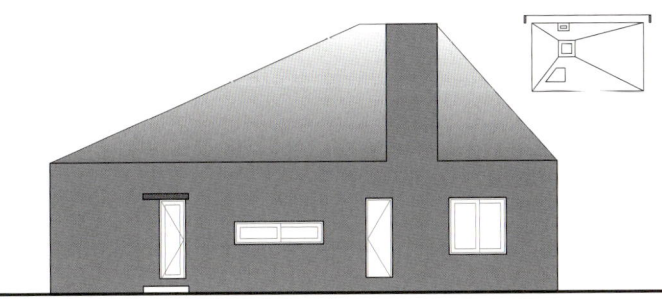

North Elevation

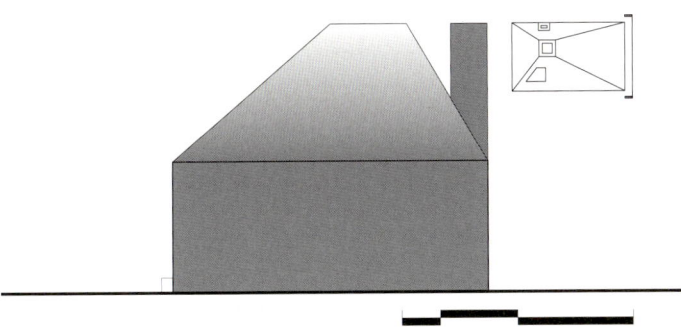

East Elevation

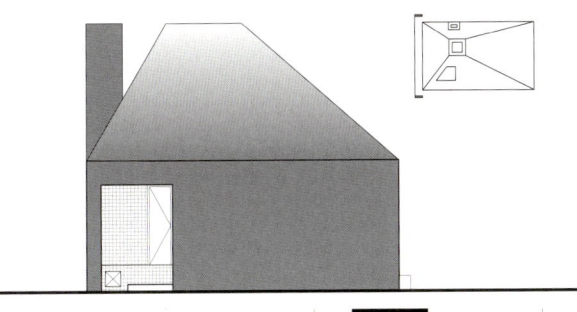

West Elevation

게르 하우스

강화도 길직리에 위치한 아담한 주택인 게르 하우스는 인천에서 사업을 하는 건축주를 위한 주말주택이다. 집터는 밭을 정리해서 주택지로 개발을 한 지역으로 완만한 경사지의 가장 높은 곳에 위치하고 있다. 건축주는 100㎡ 이하의 국민생활주택 수준으로 작게 짓기를 원했다. 오랜 아파트 생활에 익숙했던 건축주는 처음엔 잔디만 밟아도 행복할 것이라고 이야기를 했었지만 설계가 진행되면서 요구사항이 늘어나게 되었고, 그럼에도 불구하고 건축주가 지켜주길 원했던 면적과 예산으로 인해 건물을 최대한 단순화 시킬 수밖에 없었다. 외부에서 들어가는 황토 찜질방을 제외하고 모든 방과 거실을 남쪽에 배치했는데, 대면형 거실과 주방을 중심으로 좌우에 각 실을 붙여 불필요한 동선을 최소화시켰다. 계획 초기에 1층을 100㎡로 맞추고, 2층은 연면적에 포함되지 않는 다락으로 계획해 부족한 면적과 2층 테라스로의 접근을 해결하고자 했다. 하지만 다락과 1층 거실이 개방형으로 뚫려있는 것도, 다락을 통해 테라스로 나가는 것도 불가능하다는 허가권자의 규제로 다락 대신 2층으로 계획을 변경하게 되었다. 2층으로 면적이 할애되면서 건물 전체를 100㎡에 맞추기 위해 1층 면적도 줄어드는 등 전체적으로 건물의 규모를 더 줄일 수밖에 없었고 이로 인해 자칫 너무 답답하지 않을까 하는 우려를 하게 되었다. 그래서 이 주택에서는 지붕의 형태가 아주 중요했다. 박공지붕처럼 특정 방향으로 치우친 지붕이 아닌 공간의 중심이 되는 거실이 강조되도록 4면의 벽이 기울어져 중앙으로 모이고, 그곳에 천창을 설치하여 자연채광을 통해 중심공간이 더욱 빛나도록 하였다.

건축주가 정성껏 가꾼 남쪽의 정원을 바라보며 거실에 앉아 있으면 높은 층고의 개방감과 함께 천창을 통해 들어오는 빛으로 공간과 시간을 느낄 수 있다. 내부마감은 흰색으로 처리했는데, 시간에 따라 변하는 빛의 음영을 더 도드라지게 하기 위함이다.

건물의 골조가 다 올라간 다음부터 동네 사람들이 이 집을 '몽골텐트'라고 불렀다고 한다. 몽골의 전통 주거 형태인 게르는 중앙의 개구부를 중심으로 생활하는 구조로 이 집의 구조와 유사했기 때문에 이를 받아들여 게르 하우스로 이름을 지었다.

적삼목으로 포인트를 준 남쪽 면은 전통적인 한옥의 툇마루를 차용했다. 건물 형태 상 처마가 없는 지붕은 큰 창 상단의 누수 위험을 높이기 때문에 남쪽의 큰 창들은 툇마루 안쪽에 설치했고 처마가 생겨 부수적으로 계절에 따른 햇빛 유입을 조절할 수 있다. 황토방은 아궁이에 장작을 때서 방을 덥히는 고전적인 방식으로 계획되었기 때문에 화재 위험으로 적삼목 대신 목재 무늬의 타일을 붙였다. 굴뚝은 단순한 건물의 형태를 따라 단순하게 처리를 했고 밋밋한 건물의 후면에 그림자를 떨어뜨리며 포인트 역할을 하고 있다.

건축주는 처음 집을 짓겠다고 했을 때, 주말주택에 일 년에 몇 번이나 와보겠냐고 이야기를 했었으나, 완공이 되고는 평수가 훨씬 큰 아파트가 오히려 답답하다고 하며 거의 이곳에서 살고 있다. 조경에 많은 공을 들이고, 손수 가꾸면서 전원생활을 즐기고 있으며, 비만인 애완견도 살이 많이 빠졌다고 한다.

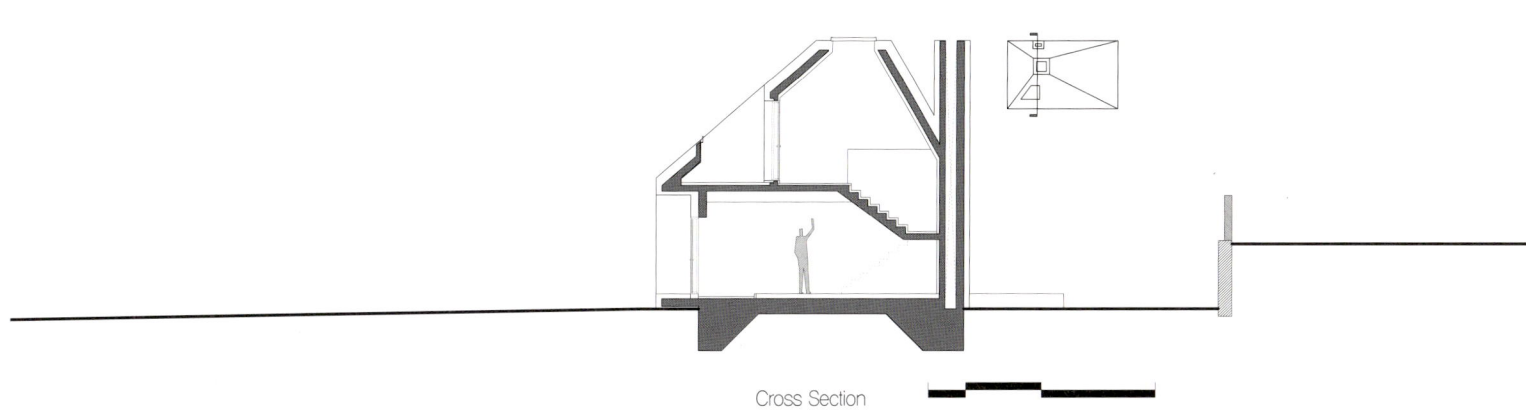
Cross Section

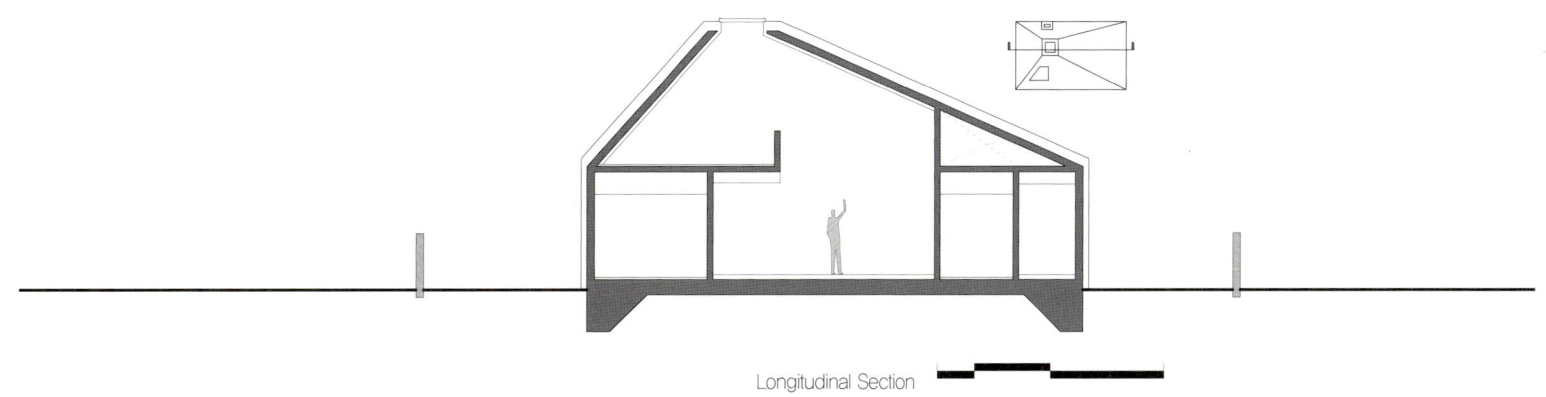
Longitudinal Section

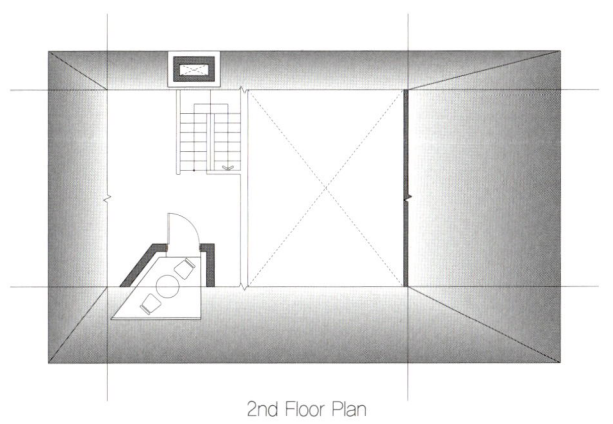
2nd Floor Plan

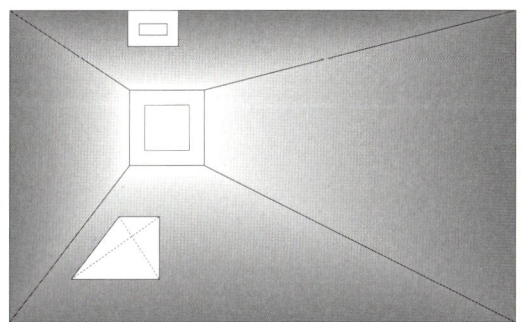
Roof Plan

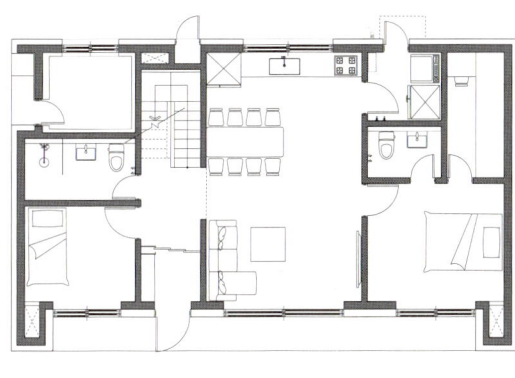
1st Floor Plan

Gaonnuri Boryeong, Korea

Gong-Yu Architecture

Architects Gong-Yu Architecture / Seongwoo Kim **Project Team** Seongjae Jeong **Location** Boryeong, Korea **Site Area** 159㎡ **Bldg. Area** 63.41㎡ **Gross Floor Area** 99.66㎡ **Bldg. Scale(Floor)** 2FL **Structure** Reinforced Concrete **Max. Height** 8.9m **Exterior Finish** T90 Glazed Brick, PVC System Windows(T22 Double-Glazed Glass) **Photographer** Houngro Yoon

설계 건축사사무소 공유 / 김성우 **설계참여자** 정성재 **위치** 충남 보령시 주교면 송학리 987 **대지면적** 159㎡ **건축면적** 63.41㎡ **연면적** 99.66㎡ **규모** 지상2층 **구조** 철근콘크리트 **최대높이** 8.9m **외부마감** T90 유약벽돌, PVC 시스템창호(T22 로이복층유리) **사진** 윤홍로

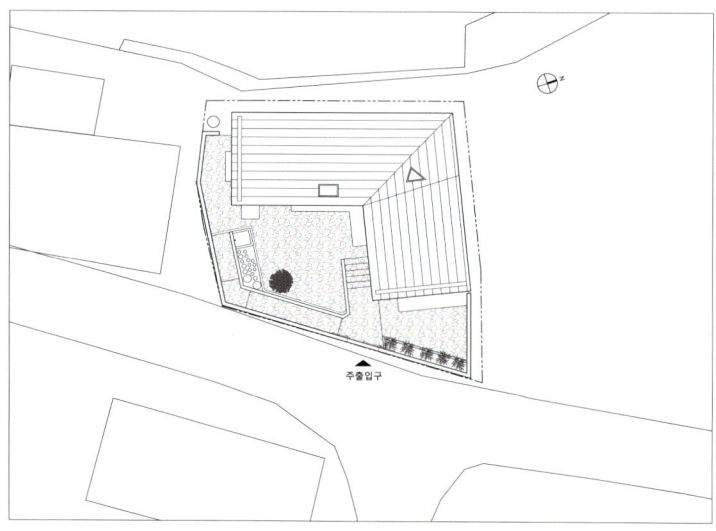

Site Plan

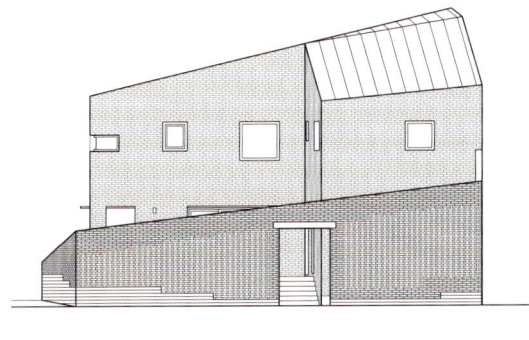
Front Elevation

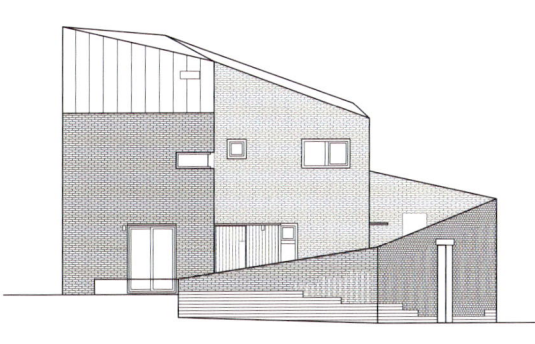
Left Elevation

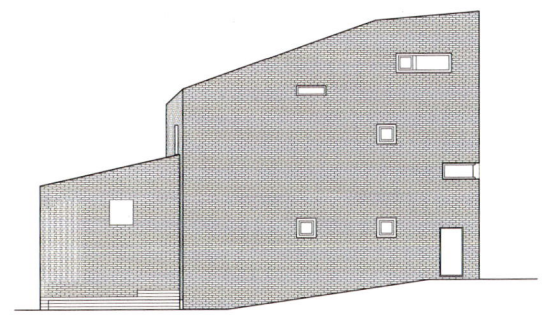
Right Elevation

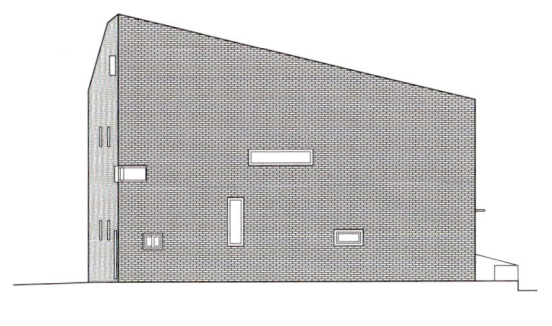
Rear Elevation

■ Filial house built for elderly parents

This house was started for elderly parents who live in an existing old house. The focus was on the daily life of living near the Boryeong beach and collecting various kinds of seafood and the situation where the parents are old. We wanted a space structure that took into consideration the movement of various kinds of seafood collected from the seashore and tidal flats in the yard and then bringing them into the kitchen, and the father with limited mobility. In addition, we wanted to provide a free space for children and grandchildren to stay when visiting, so that families can visit their parents often. Hearing these conditions in the early days, the concept of 'filial piety housing' came to mind.

■ A house like a small universe with a courtyard

The site is 48 pyeong, with a rather narrow area and irregular boundaries, and as it is a planned management area, the building-to-land ratio is limited to 40% or less and the floor area ratio is less than 100%, making it difficult to efficiently arrange houses and yards. When we looked at the land without checking these conditions, we designed a 'ㄱ' shaped courtyard house. However, the area had to be reduced due to restrictions on the building-to-land ratio. Therefore, it was finally developed into an early courtyard concept by combining the 'ㄱ'-shaped house with an open south side with a 'ㄱ'-shaped fence that gradually builds up around the courtyard and surrounds the building. The reason why we designed a house that is somewhat dignified even though it is a rural house is that privacy protection by the front pedestrian road was in mind. The wall was designed by stacking bricks with the concept of the Milky Way. The yongrong stacking makes the house invisible and allows light and wind to flow softly. In addition, when viewed from the outside, it was expected to provide various expressions depending on the position of the sun, night and day, and become a visual fun factor.

■ Space composition that melts the convenience of everyday life

The first floor is the main living space of the elderly parents, and was designed in a 'ㄱ' shape that surrounds the courtyard, and the master bedroom, bathroom, laundry, living room, kitchen, and utility rooms are arranged in a sequential flow structure considering the convenience of use. In particular, as a house adjacent to the beach, the role of a multi-purpose room is important, so it was designed with consideration for a separate movement line to be closely related to external spaces such as tap water and Jangdokdae. The main room was arranged with a separate small yard, and it was connected with a toemaru. In the central space of the 'ㄱ' structure, a dining and living space was arranged to form an open relationship with the courtyard.

■ Epilogue

A good architecture may contain a grand concept, but sometimes it is enough to contain a very small thing, a precious value. Wouldn't it be good architecture to make us reminisce, think, and stay for a while? A piece of space in our memories that we forgot. Jangdokdae and water tap… These small spatial elements occupy a corner of the yard. These spaces, made of cement rather than any expensive finishing materials, give a sense of dynamism and staying in life. The open sea and fields become my front yard. The first floor rooftop yard is eco-friendly and energy-saving by planting lawns and installing wooden decks.

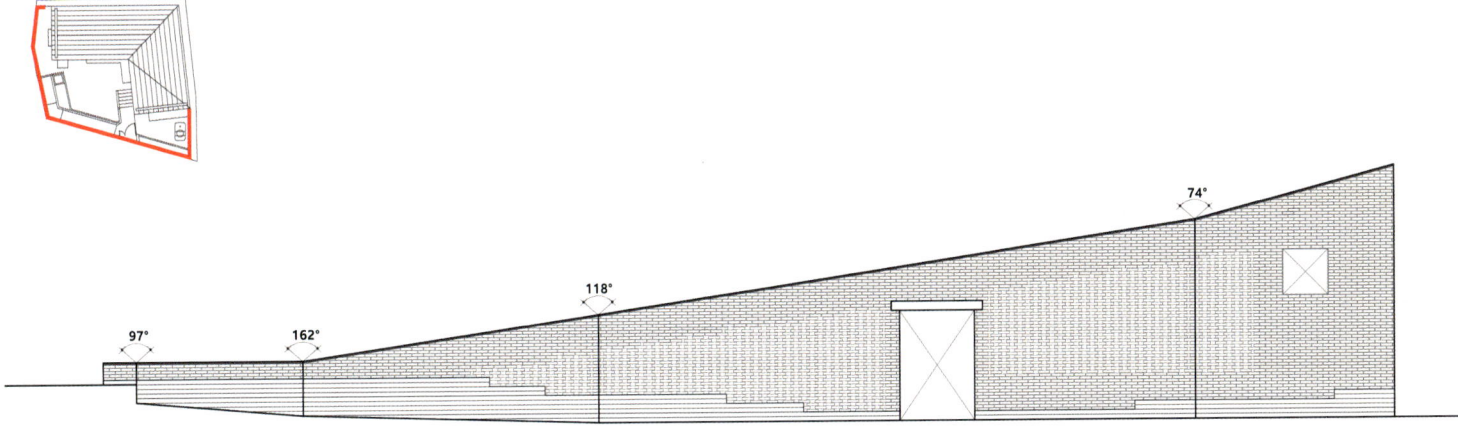

Fence Development(Milky Way Wall)

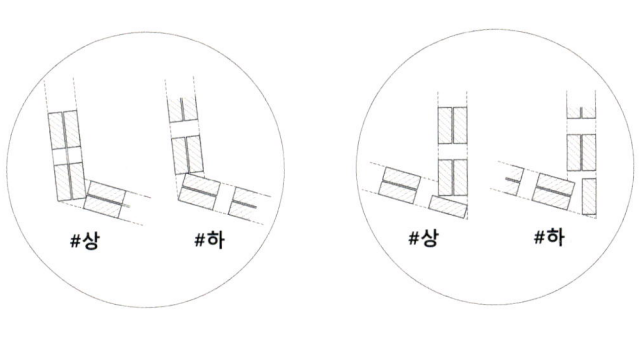

Corner Obtuse　　　　　Corner Acute Angle

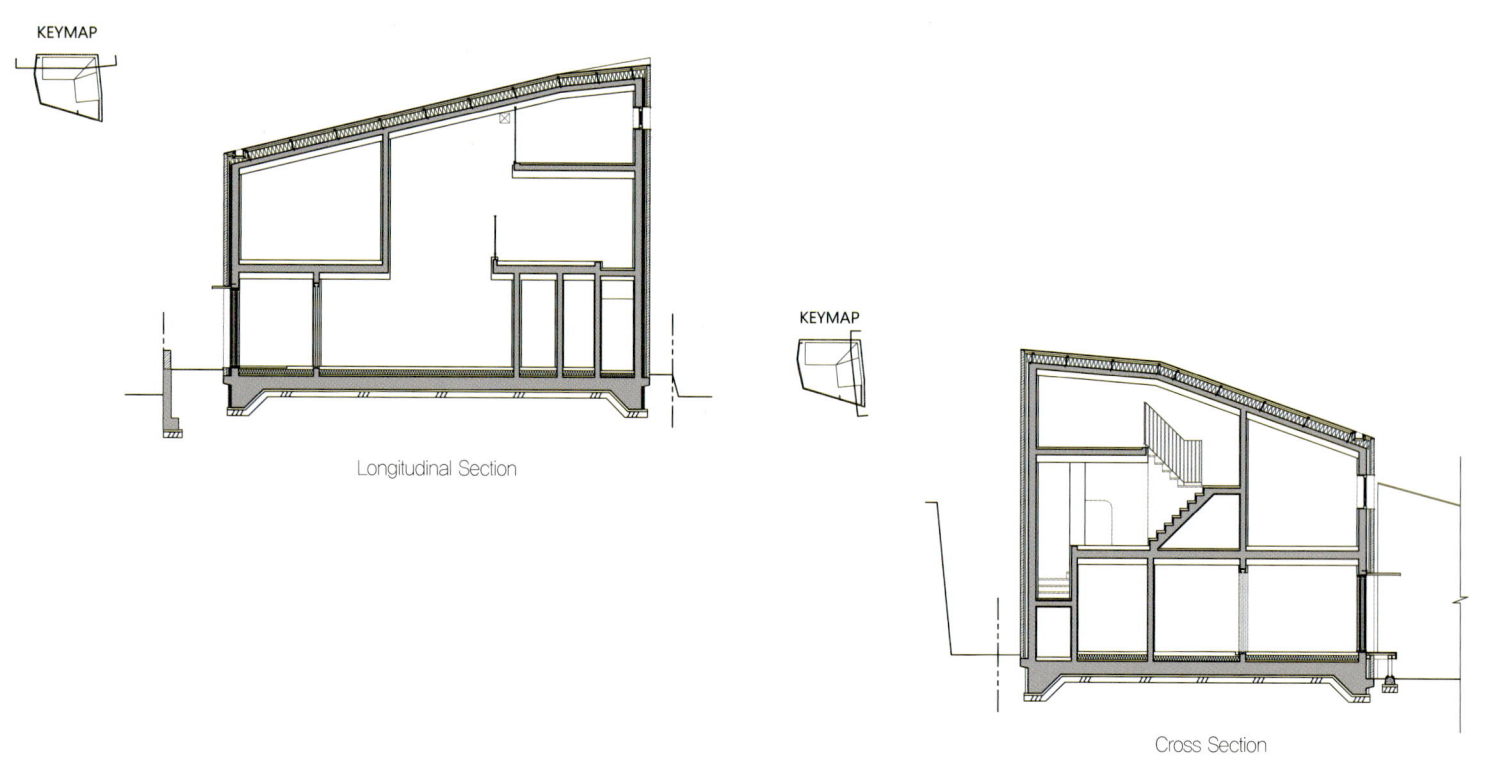

Longitudinal Section

Cross Section

가온누리

■ 노부모님을 위해 지은 효도주택

이 주택은 기존 노후 주택에 사시는 노부모님들을 위해 시작되었다. 보령 바닷가 근처에 살면서 각종 해산물을 채취하는 일상 생활과 부모님이 노령이라는 상황에 초점이 맞춰졌다. 바닷가 갯벌 등에서 채취해 온 각종 해산물을 편리하게 마당에서 손질한 뒤 주방으로 반입하는 동선 및 거동이 불편한 아버지를 배려한 공간 구조 등을 원했다. 또한 자녀들과 손주들이 방문시 머물 수 있는 여유 공간을 함께 담아 가족들이 부모님을 자주 찾아뵐 수 있기를 원했다. 초기에 이런 조건들을 듣고 '효도주택'이라는 개념이 떠올랐다.

■ 중정을 품은 소우주(小宇宙) 같은 주택

대지는 48평으로 면적이 다소 협소하고 경계는 불규칙하며, 계획관리지역이라 건폐율 40%, 용적률 100% 이하로 제한돼 있어 주택과 마당을 효율적으로 배치하기 어려운 조건이었다. 이러한 조건을 확인하지 않고 땅만 봤을 때는 'ㅁ'자 중정형 주택을 구상했다. 하지만, 건폐율 제약 때문에 면적을 줄여야 했고, 최종적으로 남쪽이 열린 'ㄱ'자형 주택에 중정을 중심으로 점차 높게 쌓아 건물을 감싸는 'ㄴ'자형 담장을 조합해 초기 중정형 개념으로 발전시켰다. 시골 주택임에도 다소 위요된 주택을 구상했던 이유는 전면 보행도로에 의한 사생활 보호를 염두에 뒀기 때문이다. 담장은 은하수 개념으로 벽돌 영롱쌓기로 설계했다. 영롱쌓기는 주택을 폐쇄적으로 보이지 않게 하고 빛과 바람이 은은하게 흐르게 한다. 또한, 외부에서 볼 때 밤과 낮, 태양 위치에 따라 다양한 표정을 제공하며 시각적 재미를 주는 요소가 되기를 기대했다. 주택 입면은 다락으로 인해 전체 비례감이 다소 높아진 아쉬움이 있지만, 단순히 입면에 장식적 요소를 주기보다 주택을 감싸는 흐름의 벽으로써 설계했다. 한편 대지 주변으로 흐르는 물길이 습한 환경을 만들어 건축지반을 높여야 했다. 이로 인해 노부모의 이동이 불편해질 수 있었다. 하지만, 이러한 단점을 은하수 담장 따라 이동이 편리한 경사로를 내고 여정의 시간을 담은 공간을 만들어 장점으로 승화시켰다. 주택을 에워싼 담장이 은하수와 같이 소우주를 형성해 아늑한 주거공간으로써 노부모에게 잔잔한 시간의 변화를 느끼게 해주고 싶었다.

■ 일상의 편리함을 녹여낸 공간구성

1층은 노부모의 주 생활 공간으로써 안마당(中庭)을 감싸는 'ㄱ'자 구조로 설계하여 안방과 욕실, 세탁실, 거실, 주방, 다용도실을 이용 편의를 고려하여 순차적 동선구조로 배치하였다. 특히 바닷가에 인접한 주택으로써 다용도실의 역할이 중요하여 수돗가, 장독대 등 외부공간과 긴밀하게 연관되도록 별도 동선을 배려하여 설계되었다. 안방은 별도의 작은 마당을 두어 배치하였고, 툇마루로 연계시켰다. 'ㄱ'자 구조의 중심공간에는 식당 겸 거실공간을 배치하여 안마당과 열린관계가 형성되도록 하였다.

2층은 자녀 및 손자 등 가족들이 방문했을 때 머물도록 'ㄱ'자 평면구조 양쪽에 각각 침실을 배치하고 가운데에 작은 가족실을 설계하였다. 그리고 주택 내부가 전체적으로 답답해지지 않도록 2층의 가족실 부분을 제외한 중간 부분을 1층과 오픈 구조로 만들어 개방감을 확보했다. 또한 자녀가 세 명이라 명절 등에 많은 가족이 모일 때를 고려한 여유 공간으로써 다락을 계획했다. 특히, 다락은 손주들이 할머니 할아버지 집에서 소중한 추억을 만드는 공간으로 요구되었던 공간이다. 집을 짓는 목적은 다양하다. 이번 프로젝트는 부모님을 위해 주택을 짓는다는 점이 이 시대에 특별한 메시지를 전하는 것 같아 완공 후에도 잔잔한 여운을 남겼다. 집을 짓는 과정에서 먼 시골길을 다닐 때, 갈 때마다 손수 정성스레 밥을 지어주던 어머님의 밥맛이 그립다. 어느 날은 현장에서 온돌 바닥에 깔 자갈을 씻고 있었는데, 쌀도 씻고 자갈도 씻는 걸 보면서 밥도 짓는 것이고, 집도 짓는 다른 말을 되새기게 되었다. 세상에 완벽한 것이 없는데 부족함이 다소 있더라도 잘 가꾸면서 삶의 여유를 느끼시길 바래본다.

■ 에필로그

좋은 건축에는 거창한 개념이 깃들여져 있을 수 있으나 때로는 아주 작은 것, 소중한 가치 하나가 담겨져 있음이 충분할때가 있다. 우리로 하여금 잠시나마 기억을 되새기고, 생각하게 하고, 머무르게 하는 것이 좋은 건축 아닐까? 우리가 잊고 지냈던 기억 속의 공간 한 조각. 장독대와 수돗가... 이 작은 공간요소들이 마당 한구석을 차지하고 있다. 그 어떤 값비싼 마감재도 아닌 시멘트로 만든 이 공간들이 삶 속에 머무름과 생동감을 준다.

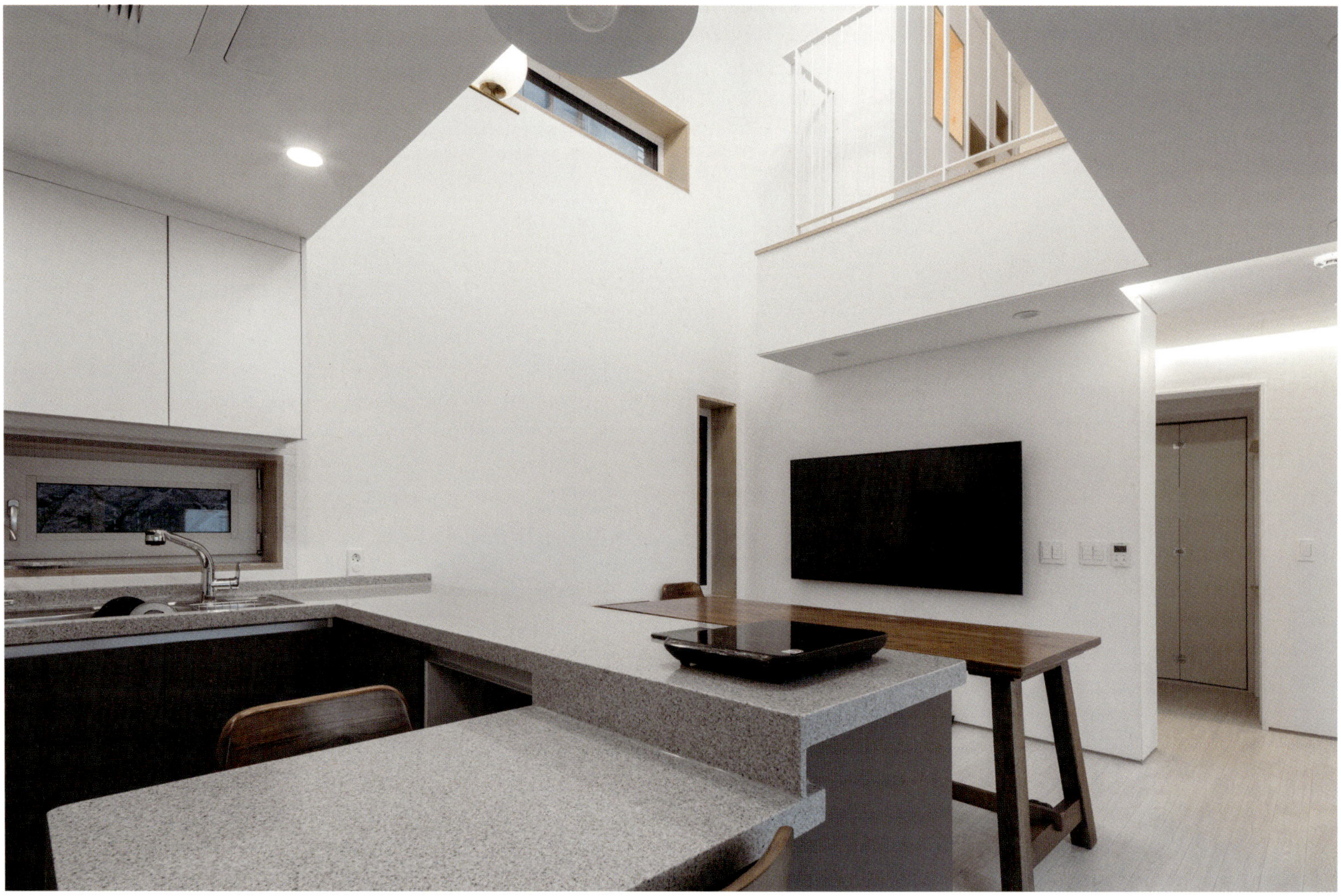

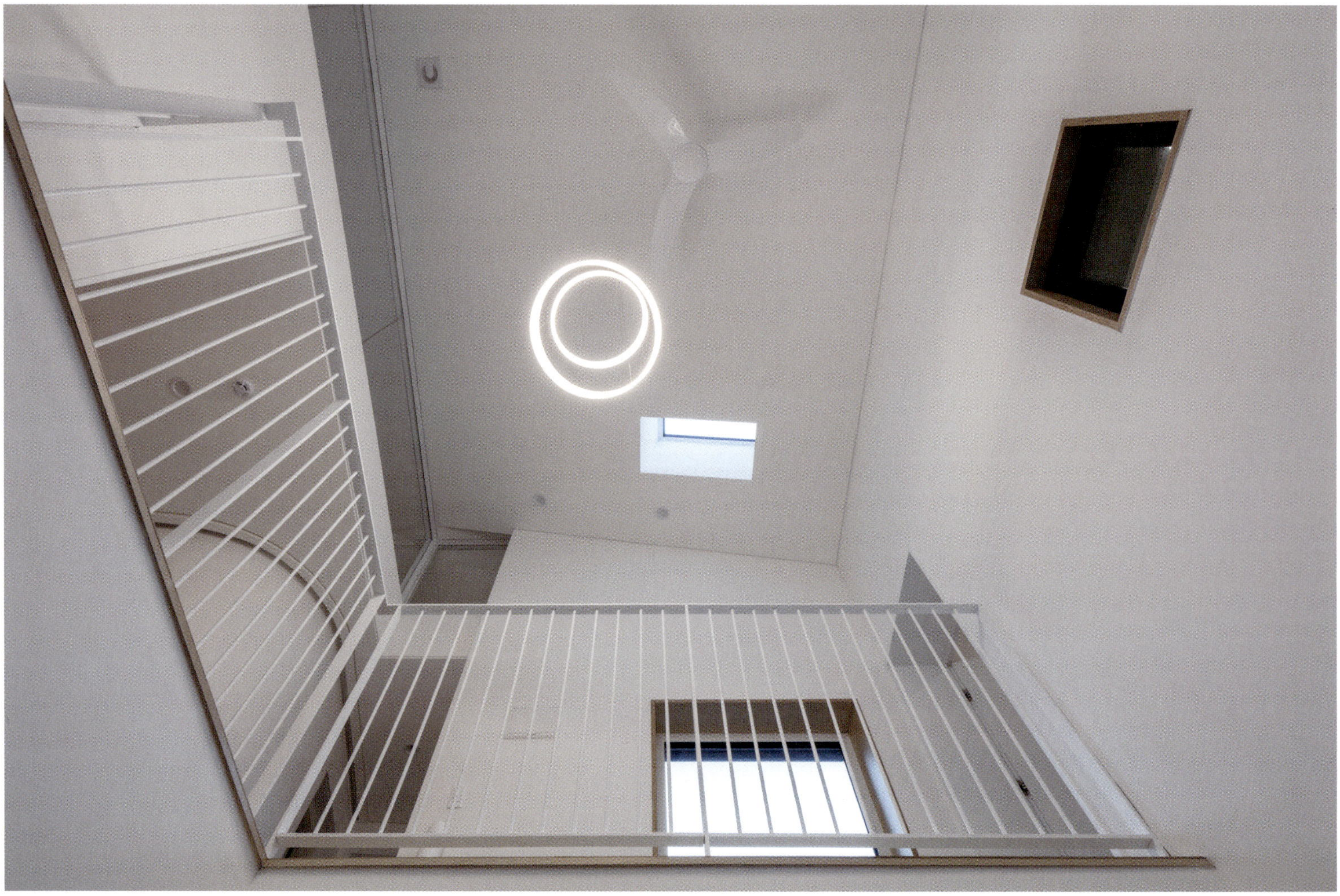

1. 1.2T PVC GUTTER(TPO, HEAT BONDING)
2. 0.5T COLOR STEEL PLATE PLAIN NAME
3. LINING ANGLE INSTALLATION(SHB175B) 175×175×12T
4. WATERPROOF PAPER(SHB140R)
5. URETHANE FOAM FILLING
6. SHB 100V DRAIN VENT
   HORIZONTAL SPACING WITHIN 60MM
7. THK1.2 GALVANIUM STEEL PLATE
   / POWDER COATING WITH SPECIFIED COLOR
8. SIDE BRICK FINISH
9. THK0.5 COLOR STEEL PLATE EXTRUSION JOINT @430
   / H:25MM
10. MOISTURE-PERMEABLE PAPER
11. 12T WATERPROOF PLYWOOD
12. 40×40×1.6T COLOR PIPE @610
13. 40×40×1.6T COLOR PIPE @1200
14. THK220 PHENOLIC FOAM (PF BOARD)
15. Ø9 STUD ANCHOR
16. 4.5T STEEL BRACKET
17. WOOD CEILING FRAME
18. THK9.5 GYPSUM BOARD 2PLY
19. APP' LAMINATED CEILING PAPER
20. THK30 EXTRUSION INSULATION PLATE NO. 1
    (INSULATION REINFORCEMENT)
21. T-SHAPED ALUMINUM MINUS MOLDING
22. THK24 BIRCH PLYWOOD(BALL FRAME) / VARNISH
23. WOOD FRAME
    / THK30 EXTRUSION INSULATION BOARD NO. 1
24. APP' PAPER WALLPAPER
25. THK7 APP' STEEL FLOOR
26. THK43 CEMENT MORTAR + HOT WATER PIPE
27. THK40 HEAT DIAPHRAGM LAYER
    (LIGHT AERATED CONCRETE)
28. THK30 EPS
29. THK210 CONCRETE SLAB
30. THK200 CONCRETE WALL
31. THK70 PHENOLIC FOAM(PF BOARD)
32. APP' BRICK(190×90×57) STUCCO STACKING
33. CONNECTION HARDWARE(SHB DS-X/SHB 85SV)
    VERTICAL 450MM / HORIZONTAL 650MM OR LESS
    DIRECT DIAMETER 4MM GALVANIZED IRON WIRE(SHB005W)
34. LINGUISTIC ANGLE (175×90×12T)
35. THK1.2 GALVANIUM STEEL PLATE
    / POWDER COATING WITH SPECIFIED COLOR
36. GALVANIZED STEEL PIPE 30×30×1.6T
37. FOLDING THE WATER CUT METAL PLATE

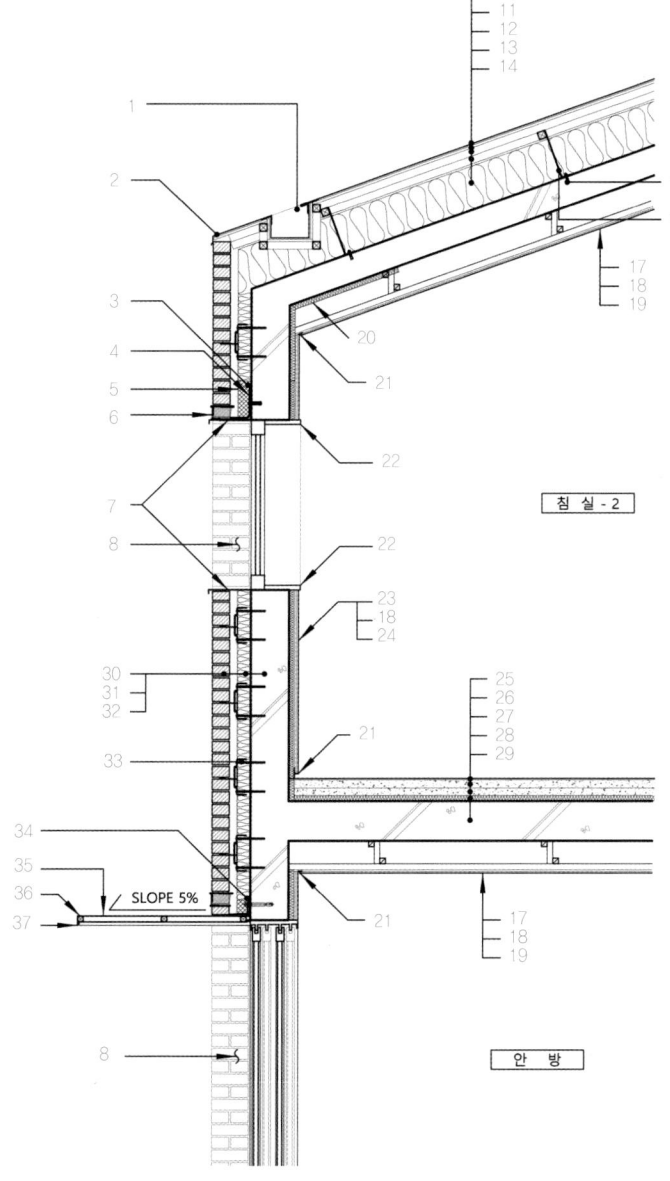

Exterior Wall Section Detail

1. 1.2T PVC거터(TPO, 열융착)
2. 0.5T 칼라강판 평이름
3. 인방용앵글설치(SHB175B) 175×175×12T
4. 방수지(SHB140R)
5. 우레탄폼충전
6. SHB 100V 배수통기구
   수평 60MM이내간격
7. THK1.2 갈바늄 강판 / 지정색 분체도장
8. 측면벽돌마감
9. THK0.5 칼라강판 돌출이음 @430 / H:25MM
10. 투습방습지
11. 12T 내수합판
12. 40×40×1.6T 컬러파이프 @610
13. 40×40×1.6T 컬러파이프 @1200
14. THK220 페놀폼(PF보드)
15. Ø9 스터드 앙카
16. 4.5T 스틸브라켓
17. 목재천장틀
18. THK9.5 석고보도 2PLY
19. 지정 합지 천장지
20. THK30 압출보온판 1호(단열보강)
21. T자 알루미늄 마이너스 몰딩
22. THK24 자작나무 합판(공틀) / 바니쉬
23. 목재틀 / THK30 압출보온판 1호
24. 지정합지벽지
25. THK7 지정강마루
26. THK43 시멘트모르타르+온수파이프
27. THK40 측열층(경량기포콘크리트)
28. THK30 EPS
29. THK210 콘크리트슬라브
30. THK200 콘크리트 벽체
31. THK70 페놀폼(PF보드)
32. 지정벽돌(190×90×57) 치장쌓기
33. 연결철물(SHB DS-X/SHB 85SV)
    수직 450MM / 수평 650MM 이내간격
    직격 4MM 아연도철선(SHB005W)
34. 인방용앵글(175×90×12T)
35. THK1.2 갈바늄 강판 / 지정색 분체도장
36. 아연도각관 30×30×1.6T
37. 물끊기 금속판접기

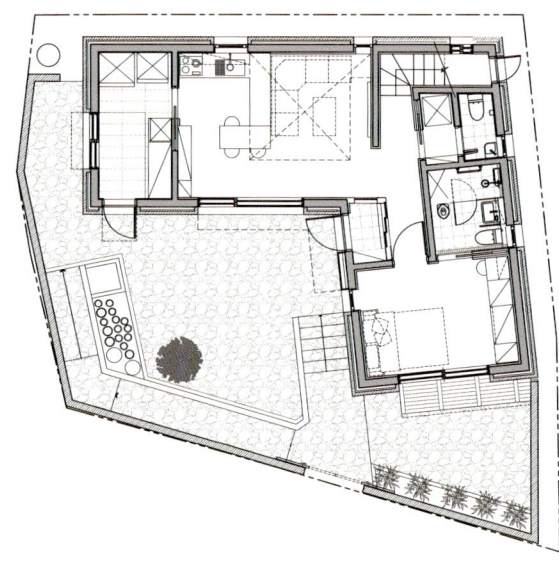

1st Floor Plan

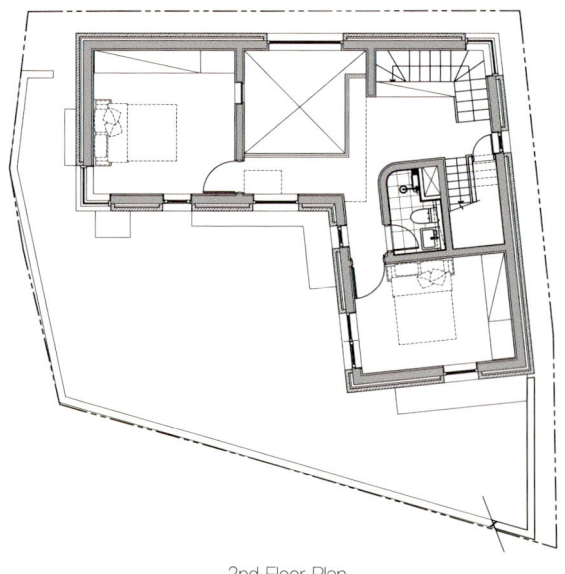

2nd Floor Plan

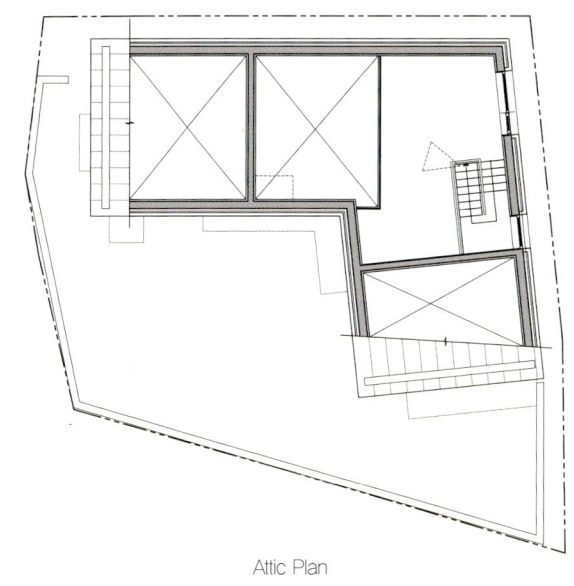

Attic Plan

Pause Inje, Korea

Architectural Design Group HAEDAM

Architects Architectural Design Group HAEDAM / Taeman An, Jeonghan Song **Project Team** Youngkyu Do, Nahyun Lee, Junghyun Lee **Location** Inje, Korea **Site Area** 998㎡ **Bldg. Area** 86.28㎡ **Bldg. Scale(Floor)** 1FL **Exterior Finish** Exterior Lime Plaster, Asphalt Shingle, Luna Wood **Client** Changcheon **Photographer** Jinbo Choi

설계 ㈜해담건축 건축사사무소 / 안태만, 송정한 **설계참여자** 도영규, 이나현, 이정현 **위치** 강원도 인제군 기린면 현리 1258-7 **대지면적** 998㎡ **건축면적** 86.28㎡ **규모** 지상1층 **외부마감** 친환경 흙미장, 바니쉬, 아스팔트쉬글, 루나우드 **발주처** 농업회사법인 창천 **사진** 최진보

Site Plan

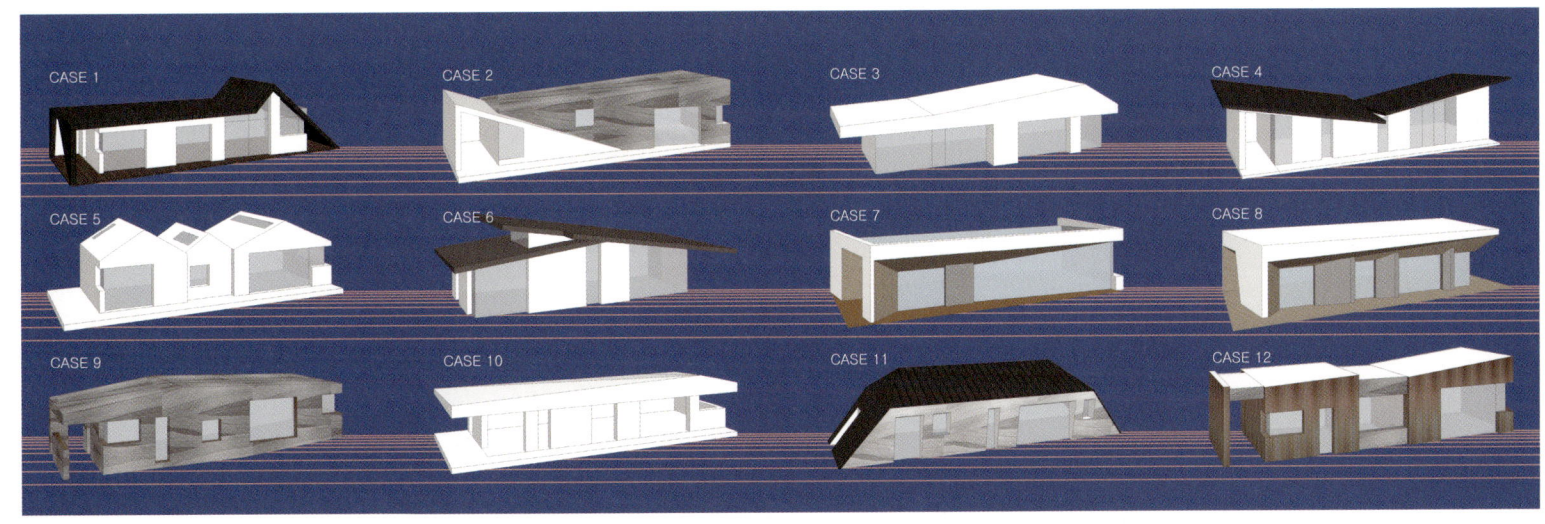

Design Development — Model Study

Front Elevation

Right Elevation

Rear Elevation

Left Elevation

Enchanted Valley, Enchanted Evergreen Forest

Eoeungol Valley in Hyeon-ri, Girin-myeon, Inje-gun is surrounded by pine trees and Korean native nut-pine trees, growing in fertile soil. Between spring and summer, pine pollen covers the ground in yellow. In winter, northern winds bring heavy and light snows in alternation. The building site was in the middle of thousands of nut-pine trees with blue-green, dark-blue, and green leaves in beautiful mountains. From the first moment we saw the site, we wanted the house to look "as if it were always there." The PAUSE had to be a harmonious part of Eoeungol and the forest.

Materials and Design: Harmony with the Neighborhood

Naturally, wood and soil were chosen as main building materials. Simple and strong structures and rafters and rough tree-bark-like earth plaster guided the design. Using the eaves of Hanok (Korean traditional house) as a motif, all outward windows were placed below 1.8 meters in height. To create a cozy atmosphere, the interior was designed to lead people to lower their heads or sit in order to fully enjoy the views of the surrounding valley and forest. To join modern convenience with Hanok-style elegance, we exposed many of the wooden structures and put lighting between the structures, while keeping the interior simple. Simplicity was our overriding principle. The rest of its identity as a house was left for its residents to fill in.

We mimicked the texture of Korean nut-pine tree bark for its exterior walls. It took two weeks to achieve the exact look that we wanted. We changed its color twice until it seamlessly blended in with the enchanted valley and forest. Finding harmony between the house and the neighborhood was challenging. The solution came from paying attention to colors and heights to suit the surroundings. This experience reminded us that our role as architects is finding ways to realize clients' vision.

Change of Identity: A Home to a Guesthouse

Initially, the house was going to be the home of an elderly couple. Visits from their children and friends during construction led to a lot of discussions about the house. Eventually, the couple decided to make it a guesthouse where people come to stay and relax in nature. Although a house is built with a specific purpose in mind, it evolves with the people who use it.

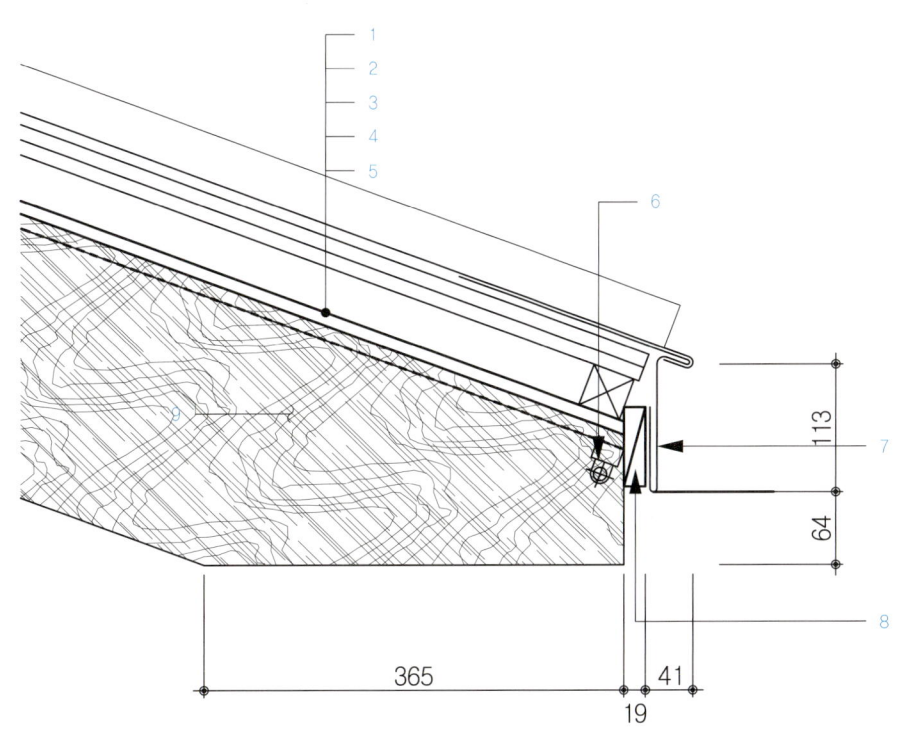

1. ASPHALT DOUBLE SHINGLE BLACK
2. WATERPROOF RESIN
3. 12MM WATERPROOF PLYWOOD
4. ㅁ-40×40×1.6 RUST PREVENTATIVE SQUARE PIPE
5. 11MM OSB PLYWOOD
6. LED T5 SINGLE LIGHT / FOR EAVES
7. COLOR STEEL SHEET / AL ALLOY
   BUTT END WATER DRIP
   FOR ICICLE FORMATION, NO EXCELLENT GUTTERS
8. APP' VARNISH ON 19×70 WOOD VENEER
9. APP' VARNISH ON RAFTERS 38×235 @610 J-GRADE

1. 아스팔트2중쉰글 블랙
2. 투습방수지
3. 12MM 내수합판
4. ㅁ-40×40×1.6 방청각파이프
5. 11MM OSB 합판
6. LED T5 단조명 / 처마용
7. 칼라강판 / AL합금
   마구리 물끊기
   고드름 맺히기용. 우수거터없음
8. 19×70 목재 덧판 위 지정바니쉬
9. 서까래 38×235 @610 J-GRADE 위 지정바니쉬

Eaves Section Detail

1. RAFTER : 2" × 10"
2. RIDGE BOARD : 2" × 12"
3. HIP RAFTER : 2" × 12"
4. CEILING JOIST : 2" × 10"
5. STUD : 2" × 6"
6. PSL : 4" × 10"
7. HEADER : 4 – 2" × 10"
8. THK300 MAT

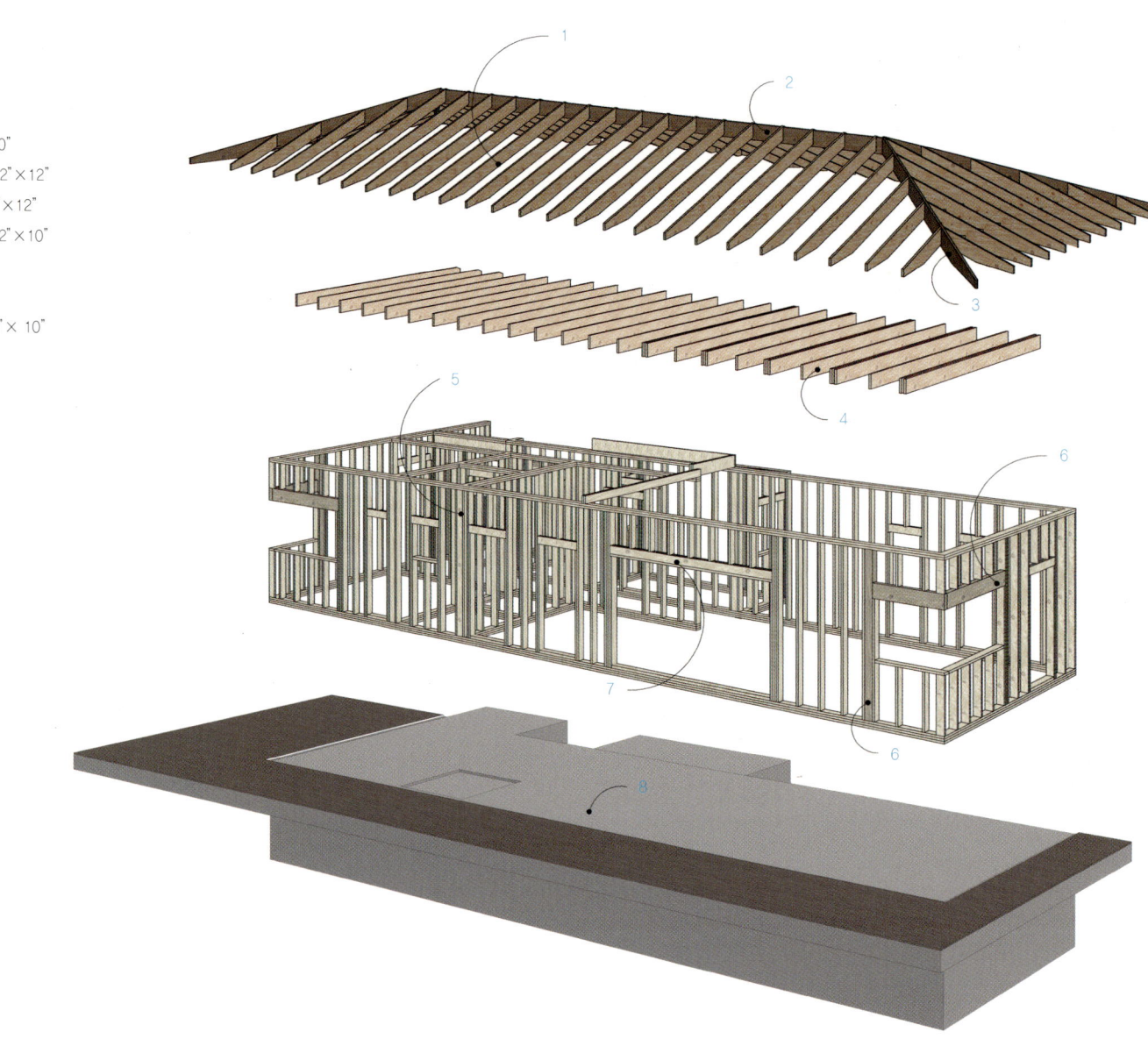

Structure + Space Analysis Isometric

파우재

서로매혹의 계곡, 매혹의 상록수림

소나무와 잣나무가 많고 흙이 좋은 인제군 기린면 현리의 어은골 계곡. 봄과 여름 사이에는 송화가루가 노랗게 쌓이고, 겨울엔 삵바람이 지나가면서 묵직한 눈과 쌀가루 눈이 번갈아 가면서 내린다. 청록색이면서 남색이고 녹색이기도한 잎을 지닌 수천그루의 잣나무, 소나무들의 뒷산. 그들을 처음본 순간 이 집에 대한 바램은 '처음부터 거기에 있었던 것처럼'이었다. 파우재가 어은골과 잣나무숲의 일부가 되기를 바랐다.

재료와 구법구현, 지역성과 장소, 같이 풀어내기

당연스럽게도 집을 짓는 재료는 나무와 흙이 되었다. 소박하면서도 힘있는 뼈대와 서까래, 거칠면서 자연스런 질감을 느낄수 있는 잣나무 껍질같은 흙미장이 기본이 되었다. 한옥의 처마선이 기본이 되어 창문의 높이는 전부 1.8m 이하로 낮추었다. 고개를 조금 숙이거나 앉아야만 계곡과 숲을 볼 수 있게 하여 집의 위압감을 덜어내려 하였다. 요즘 주택의 편의성은 유지하고 한옥과 같이 목조주택의 맛을 살리기 위해 실내에서 목구조 프레임을 최대한 노출시키면서, 구조목을 조명의 기본으로 사용하고, 모양은 최대한 단순하게 하였다. 소박하고 단졸함을 기본으로 하고 집으로서의 정체성의 나머지는 살아가는 가족에게 맡겼다.

잣나무 껍질의 미감을 흉내내어, 흙미장의 깊이와 모양을 2주일 동안 수정하면서 만들어냈다. 색깔은 시공 시행착오와 결과의 잘못을 건축가인 우리가 인정하고 두 번의 수정 과정을 거쳐서 매혹의 계곡과 숲에 누가 되지않게 결과가 나왔다. 지역성은 재료와 구법에서, 장소에 대한 고민은 의외로 색깔과 바라보는 높이에서 근접점을 찾았다. 건축가는 풀어가는 사람이지 강요하는 존재는 아닌 것 같음을 새삼 느꼈다.

처음엔 집, 지금은 스테이. 정체성의 변화

처음엔 두 노부부가 살아가는 집이 건축의 목적이였다. 지어지는 과정에서 건축주 자녀분과의 대화와 교감이 많이 이루어지고, 지인들이 다녀가면서 사용의 목적이 바뀌기 시작했다. 이제는 많은 사람들이 묵고가는 스테이(민박)가 되었다. 집이란 지역성, 장소, 재료와 구법이 잘 혼합되어 지어지지만 물건만이 아니기에 이용하는 사람에 따라서 다른 무엇이 되는 것 같다.

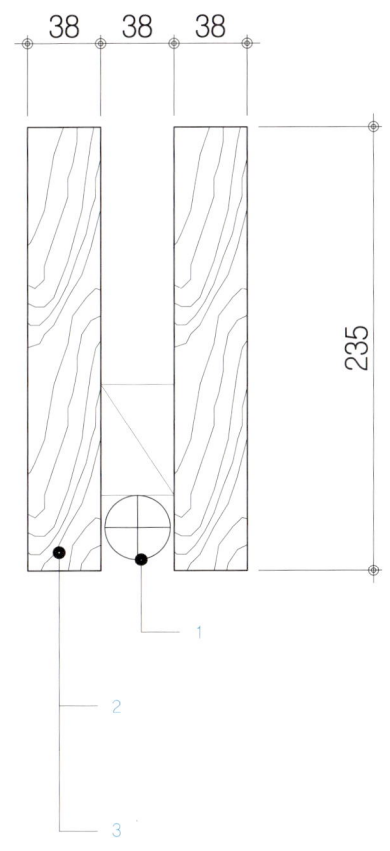

1. LED T5 WHITE, PLACED IN LIGHT BULB-COLORED ALTERNATING ROWS
2. CEILING JOISTS 38×235 @610
   STRUCTURAL BEAM / THE ROLE OF GIRDER IN HANOK
   STRUCTURAL MATERIAL AND LIGHTING ROLE / NO SEPARATE LIGHTING FOR LIVING ROOM
3. PAINTED WITH WATER-BASED VARNISH TRANSPARENT

1. LED T5 주백색, 전구색 교차열로 배치함
2. 천장장선 38×235 @610
   구조보 / 한옥의 대들보 역할
   구조재이자 조명 역할 / 거실 별도 조명 없음
3. 수성바니쉬 투명으로 도장

Joist and Lighting Section Detail

1. BEDROOM
2. BATHROOM
3. LIVING ROOM / KITCHEN
4. CORRIDOR

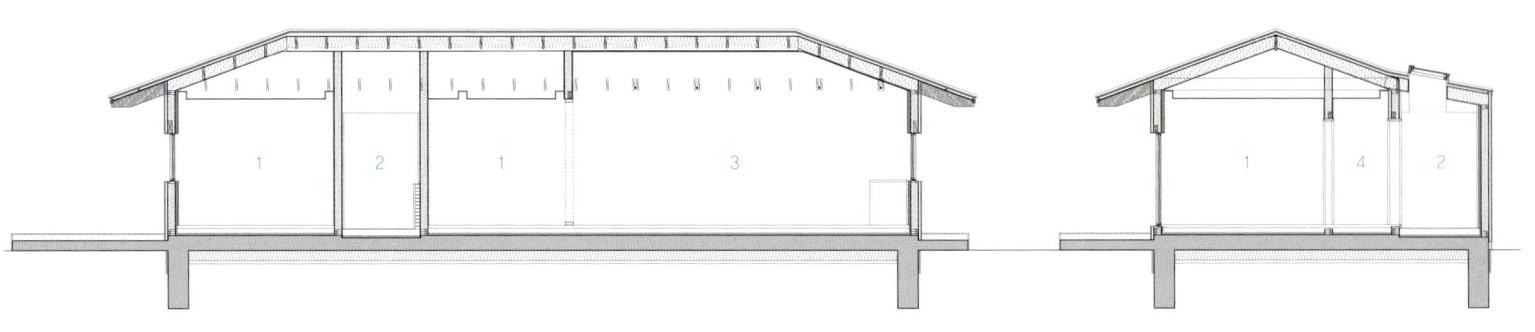

Cross Section   Longitudinal Section

1. DECK 5. POWDER ROOM
2. BOILER ROOM 6. BATHROOM
3. ENTRANCE 7. LIVING ROOM
4. BEDROOM 8. KITCHEN

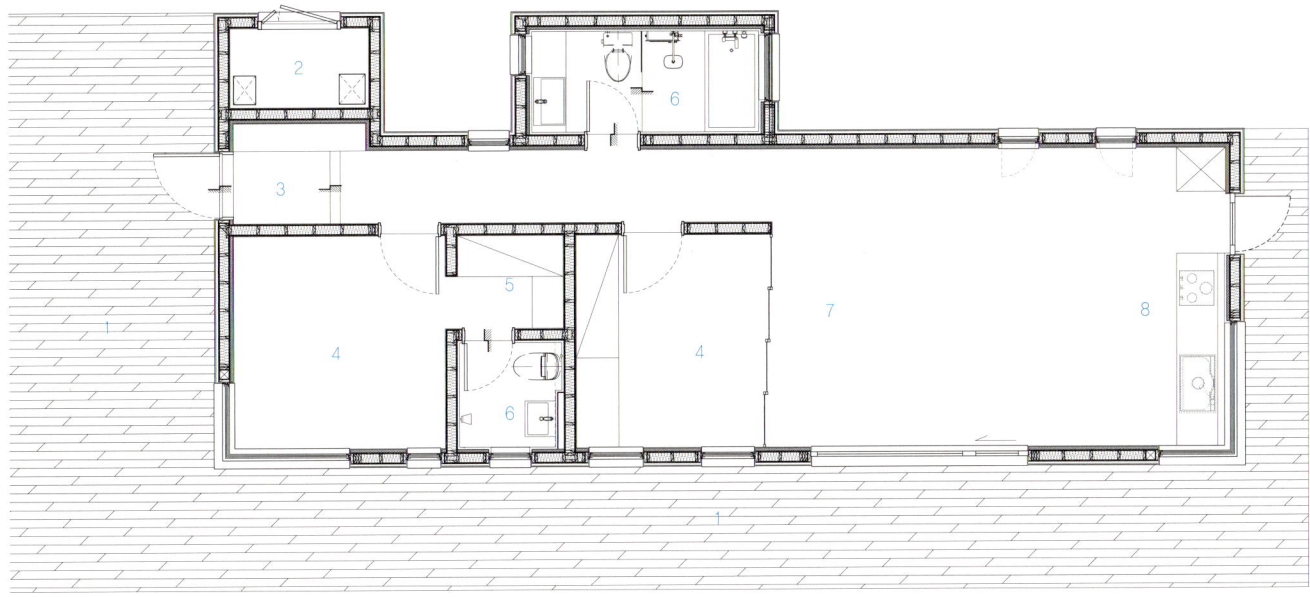

1st Floor Plan

Hokepos Seoul, Korea

NAOI + PARTNERS

Architects NAOI + PARTNERS / KATSUTOSHI NAOI **Project Team** Minyong Park **Location** Seoul, Korea **Site Area** 281㎡ **Bldg. Area** 79.04㎡ **Gross Floor Area** 148.3㎡ **Bldg. Scale(Floor)** 2FL **Structure** Reinforced Concrete **Max. Height** 8.96m **Exterior Finish** Brick **Photographer** Seongchul Kim

설계 나오이플러스파트너스 / 나오이 카츠토시 **설계참여자** 박민용 **위치** 서울시 성북구 **대지면적** 281㎡ **건축면적** 79.04㎡ **연면적** 148.3㎡ **규모** 지상2층 **구조** 철근콘크리트 **최고높이** 8.96m **외부마감** 치장벽돌 **사진** 김성철

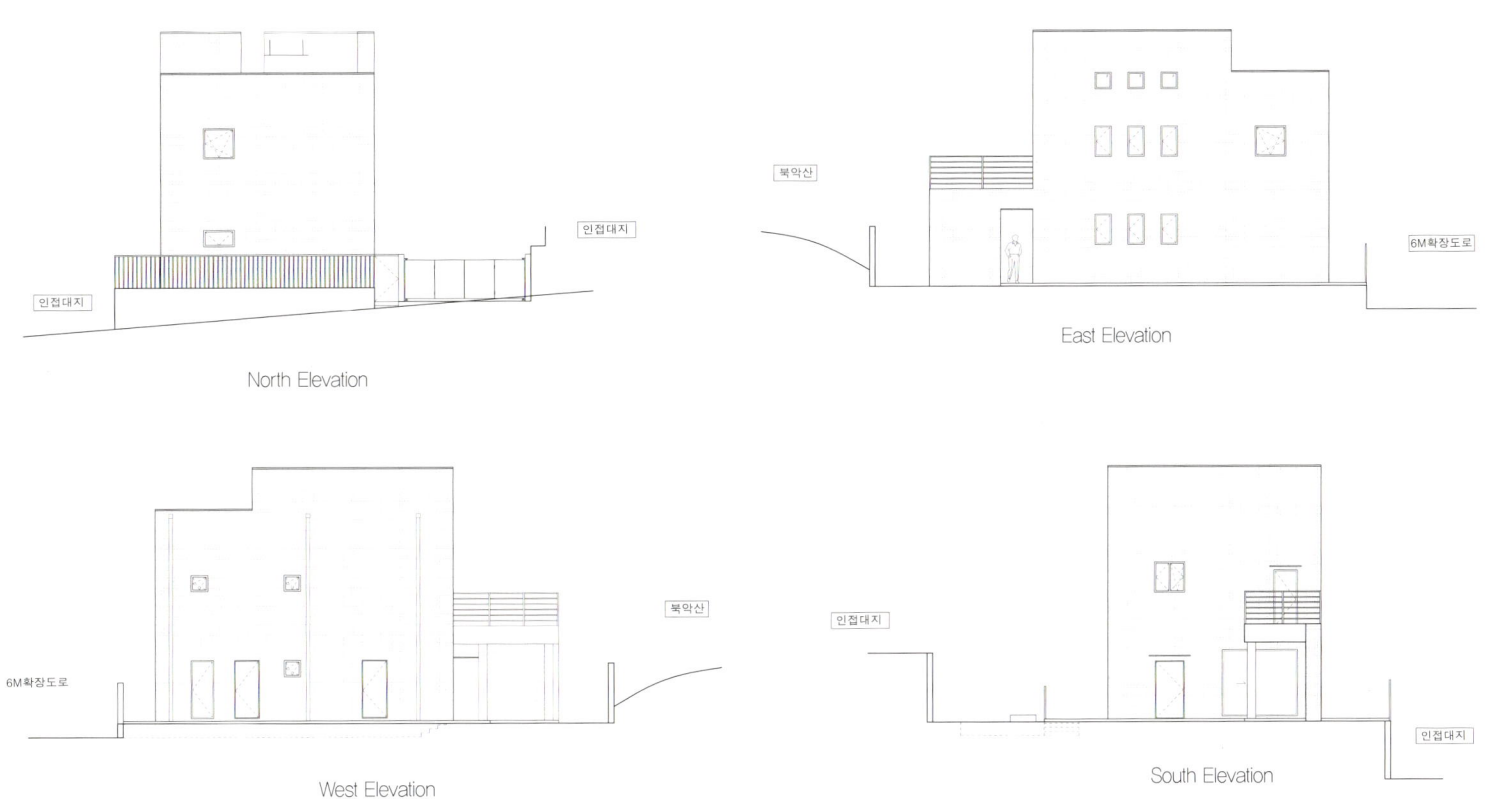

North Elevation

East Elevation

West Elevation

South Elevation

Meeting with the owner and planning the project
(Building plan, how we met, consultation process and project progress, including client requirements)
The client, who had always been interested in Italian culture and language, came to us with a site in Jeongneung-dong, where a house that had not been used for over 10 years was located. The project began when the client, who wanted to build a good house for her family, showed the cover of a CD album with photos of rural farmhouses in Europe and the formidable condition that there are about 3,000 books.

Site conditions and location
(Location Conditions and Reason for Site Selection)
The site was located almost at the top of Bukaksan Mountain along the narrow alleyways. To the south, it faced the dense Bukaksan forest, and it was a site that had a close relationship with the road. The road facing the road looks quiet, but there were quite a few people coming and going to the temple and the entrance to the hiking trail. And I thought that the appearance of the two-story house on the given site would be recognized in various ways as I climbed up along the winding alleys, and I tried to capture this point in the elevation and volume.

House layout
(Housing arrangement considering the shape of the site and the surrounding landscape)
I thought it was necessary to protect privacy from the road frequented by outsiders, so I decided to plan the entrance to the back of the building (south-facing) according to the owner's request. On the north-facing road, it was decided to minimize the windows and to receive the necessary sunlight from the east and the south. The client wanted to build a two-story building as large as the size of the old abandoned house, so the layout of the existing building was used to the maximum.

Elevation plan
(Elevation design concept, visual (formal beauty) and environmental design elements)
The owner wanted to use bricks, so he wanted to show the character of a wall made of bricks. The color and texture of the bricks were chosen not to deviate significantly from the surrounding environment, and the windows were planned to reveal the character of a wall through certain sizes and rules.

Floor plan and features of each room
(Flat plan and space arrangement considering the client's needs and lifestyle)
The first floor was arranged as a common space for family members and the second floor followed the basic layout of placing a private space. However, since I own a lot of books, I had to satisfy the requirements by using a variety of spaces rather than a general library. The space facing Bukaksan was planned to take the relationship between the inside and outside space more actively by using a space called 'loggia' rather than a typical deck or terrace. Loggia: As a spatial form that appears frequently in Italian country houses, it refers to the space of the corridor covered with a roof, and plays a role as a medium to connect the relationship between the interior and exterior spaces.

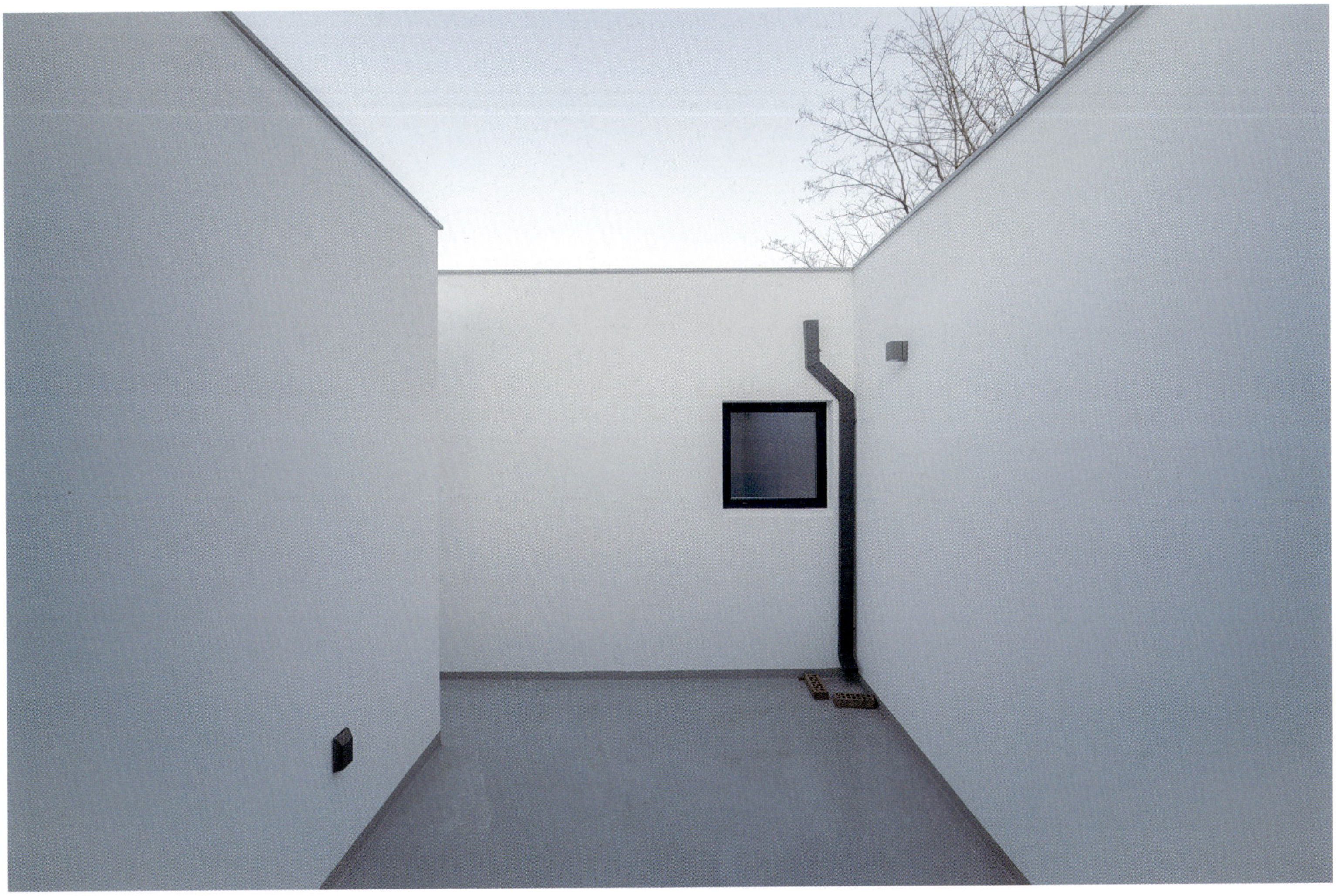

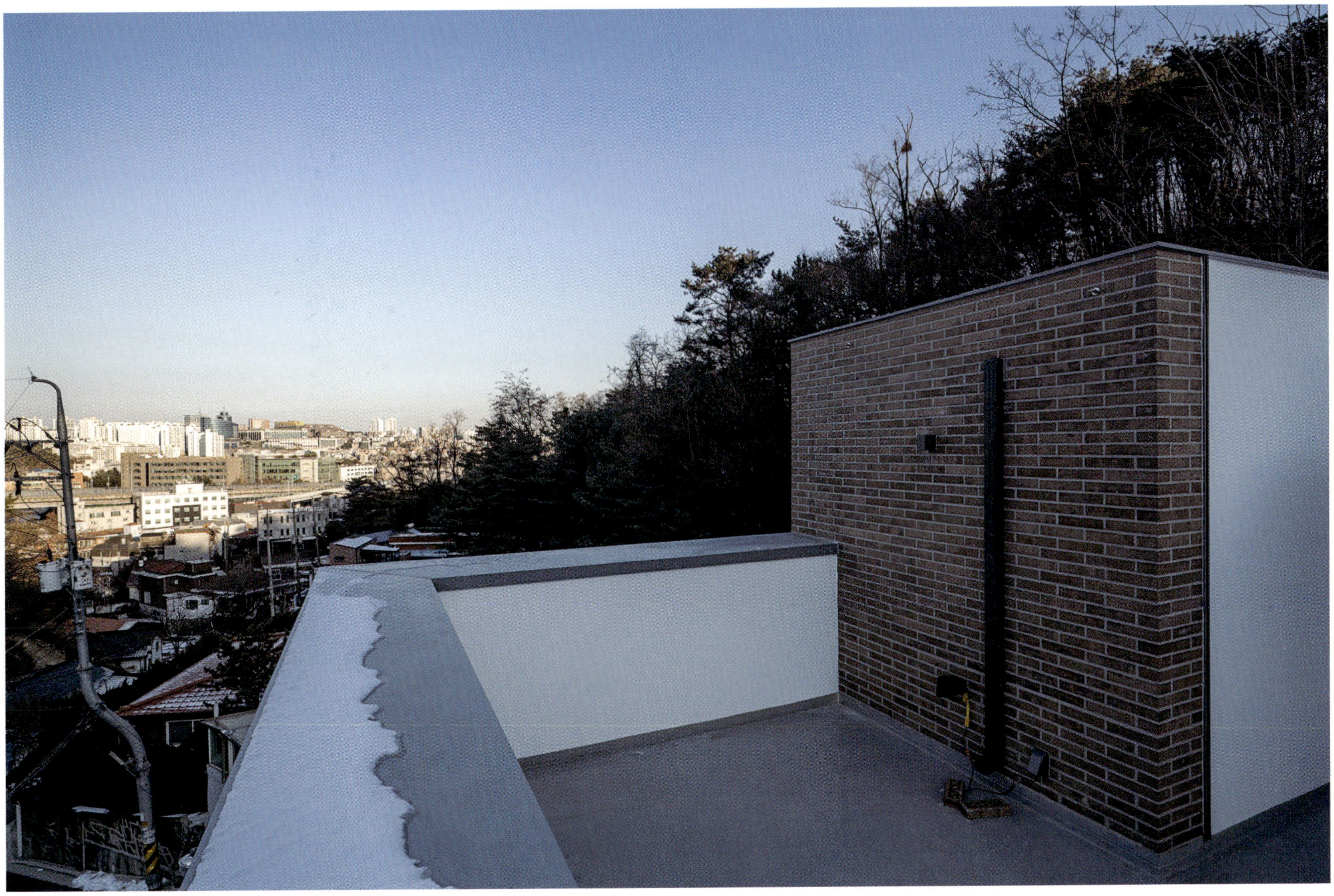

호케포스

■ 건축주와의 만남 및 프로젝트 계획
(건축 계획, 어떻게 만났는지, 상담 과정 및 건축주 요구 사항 등 프로젝트 진행 과정)
- 이탈리아의 문화와 언어에 평소 관심이 많았던 건축주는 10년 넘도록 사용하지 않았던 집이 위치한 정릉동의 한 부지를 가지고 찾아주셨다. 가족들을 위해 좋은 집을 짓고 싶다는 건축주는 책이 3000여권 정도 있다는 만만치 않은 조건과 유럽의 시골 농가주택의 사진이 담긴 CD 음반의 표지를 보여주면서 프로젝트는 시작되었다.

■ 대지 여건 및 입지
(입지 조건과 대지 선택 이유)
- 대지는 좁은 골목길을 따라 북악산 자락을 올라가다 보면 거의 꼭대기에 위치하고 있었다. 남쪽에는 울창한 북악산 숲과 마주하고 있었고, 도로와 가깝게 관계를 맺고 있는 대지였다. 마주하고 있는 도로는 한적해 보이지만 절과 등산로 입구 등으로 꽤 많은 사람이 왕래하고 있었다. 그리고 고불고불한 골목길을 따라 올라오다 보면 주어진 대지의 2층집의 모습이 다양한 모습으로 인식될 것이라고 생각했고, 이러한 점을 입면과 볼륨에 잘 담아내려고 하였다.

■ 주택 배치
(대지 형태와 주변 경관 고려한 주택 배치)
- 외부인의 통행이 잦은 도로로부터 프라이버시를 지켜야 한다고 생각하였고, 건축주의 요구대로 현관 입구를 건물 뒤쪽(남향)으로 계획하기로 하였다. 북향인 도로에는 창문을 최소화하고 동향과 남향에서 필요한 햇빛을 받기로 하였다. 건축주는 기존에 있던 오래된 폐가의 크기만큼 2층으로 짓고 싶어했기에 기존건물의 배치를 최대한 이용했다.

■ 입면 계획
(입면 디자인 콘셉트, 시각(조형미) 및 환경 고려한 디자인 요소)
- 건축주는 벽돌을 사용하는 것을 원하였기 때문에 벽돌로 쌓아서 만드는 벽이라는 성격을 잘 드러내고 싶었다. 벽돌의 색상과 질감은 주변의 환경과 크게 벗어나지 않는 것으로 선택하였고, 창문들은 일정한 크기와 규칙들을 통해 벽이라는 성격을 잘 드러내도록 계획하였다.

■ 평면 계획 및 각 실별 특징
(건축주 요구와 라이프 스타일 고려한 평면 계획과 공간 배치)
- 1층은 가족간의 공용공간으로 배치하고 2층은 사적인 공간을 두는 기본적인 배치를 따랐다. 하지만 책을 많이 소유하고 있어서 일반적인 서재가 아닌 다양한 공간을 이용해 요구조건을 만족해야 하는 상황이었다. 북악산과 마주하고 있는 공간은 일반적인 데크나 테라스가 아닌 '로지아'라는 공간을 이용하여 내외부 공간의 관계를 더욱 적극적으로 가져가려고 계획하였다.

*로지아: 이탈리아 전원주택에서 많이 나타나는 공간적 형태로서 지붕으로 덮인 회랑의 공간을 이야기하며 내외부 공간의 관계를 잘 이어주는 매개체의 역할을 한다.

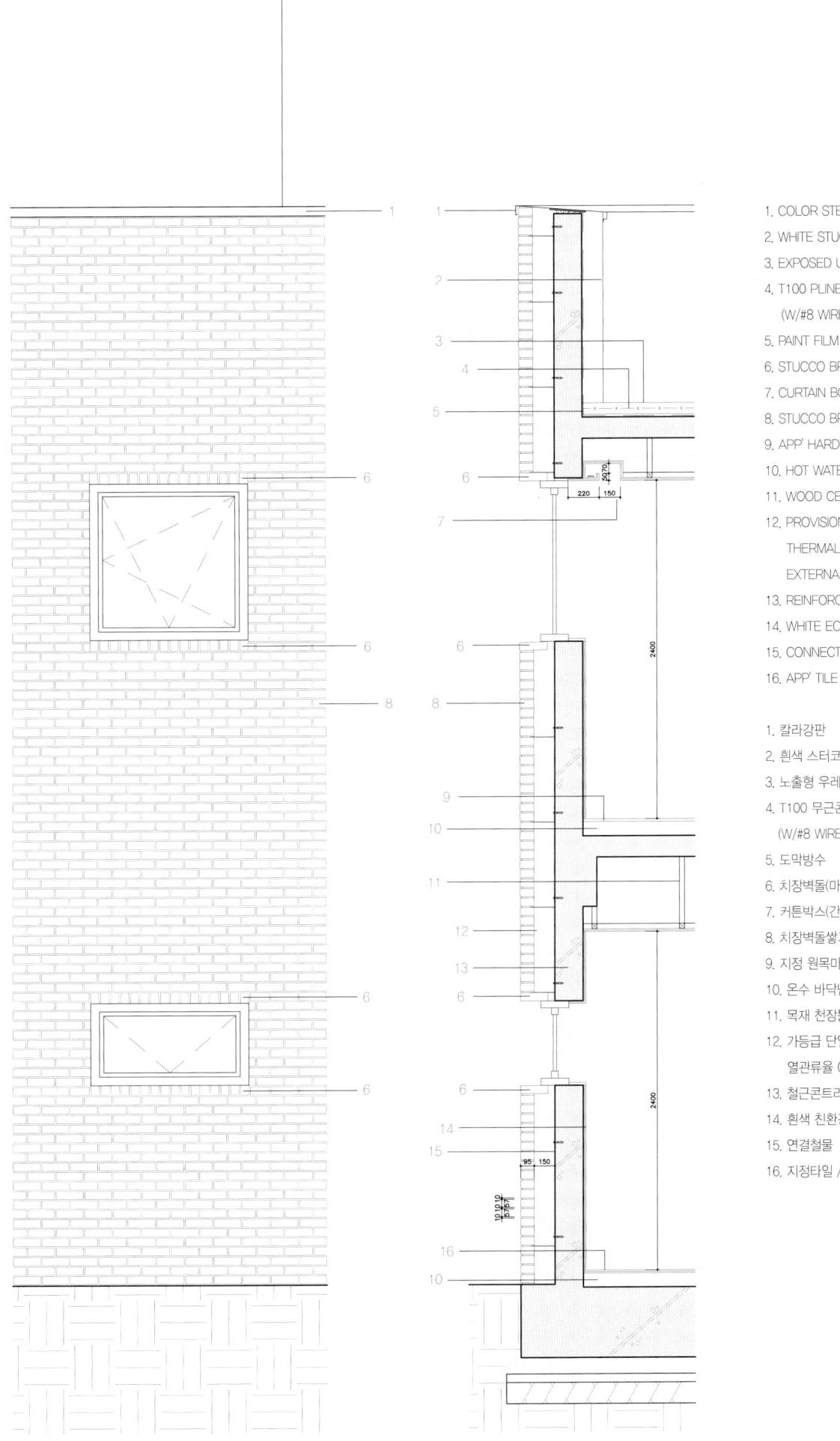

1. COLOR STEEL PLATE
2. WHITE STUCCO(WIRE MESH #8-150×150)
3. EXPOSED URETHANE WATERPROOFING
4. T100 PLINE CONCRETE IRON TROWEL FINISH
   (W/#8 WIREMESJ @100×100)
5. PAINT FILM WATERPROOFING
6. STUCCO BRICKS(MAGURI-MYEON ERECTION)
7. CURTAIN BOX(INDIRECT LIGHTING)
8. STUCCO BRICKWORK 240×90×50
9. APP' HARDWOOD FLOOR FINISH
10. HOT WATER UNDERFLOOR HEATING
11. WOOD CEILING FRAME
12. PROVISIONAL GRADE INSULATION, 135MM
    THERMAL TRANSMITTANCE LESS THAN 0.26,
    EXTERNAL INSULATION
13. REINFORCED CONCRETE, 200MM
14. WHITE ECO-FRIENDLY W.P 3 TIMES
15. CONNECTING HARDWARE
16. APP' TILE / MORTAR

1. 칼라강판
2. 흰색 스터코(와이어매쉬#8-150×150)
3. 노출형 우레판 방수
4. T100 무근콘크리트 쇠흙손마감
   (W/#8 WIREMESJ @100×100)
5. 도막방수
6. 치장벽돌(마구리면 세워쌓기)
7. 커튼박스(간접조명)
8. 치장벽돌쌓기 240×90×50
9. 지정 원목마루 마감
10. 온수 바닥난방
11. 목재 천장틀
12. 가등급 단열재, 135MM
    열관류율 0.26이하, 외단열
13. 철근콘트리트, 200MM
14. 흰색 친환경 W.P 3회
15. 연결철물
16. 지정타일 / 모르타르

Section Detail

Cross Section

Longitudinal Section

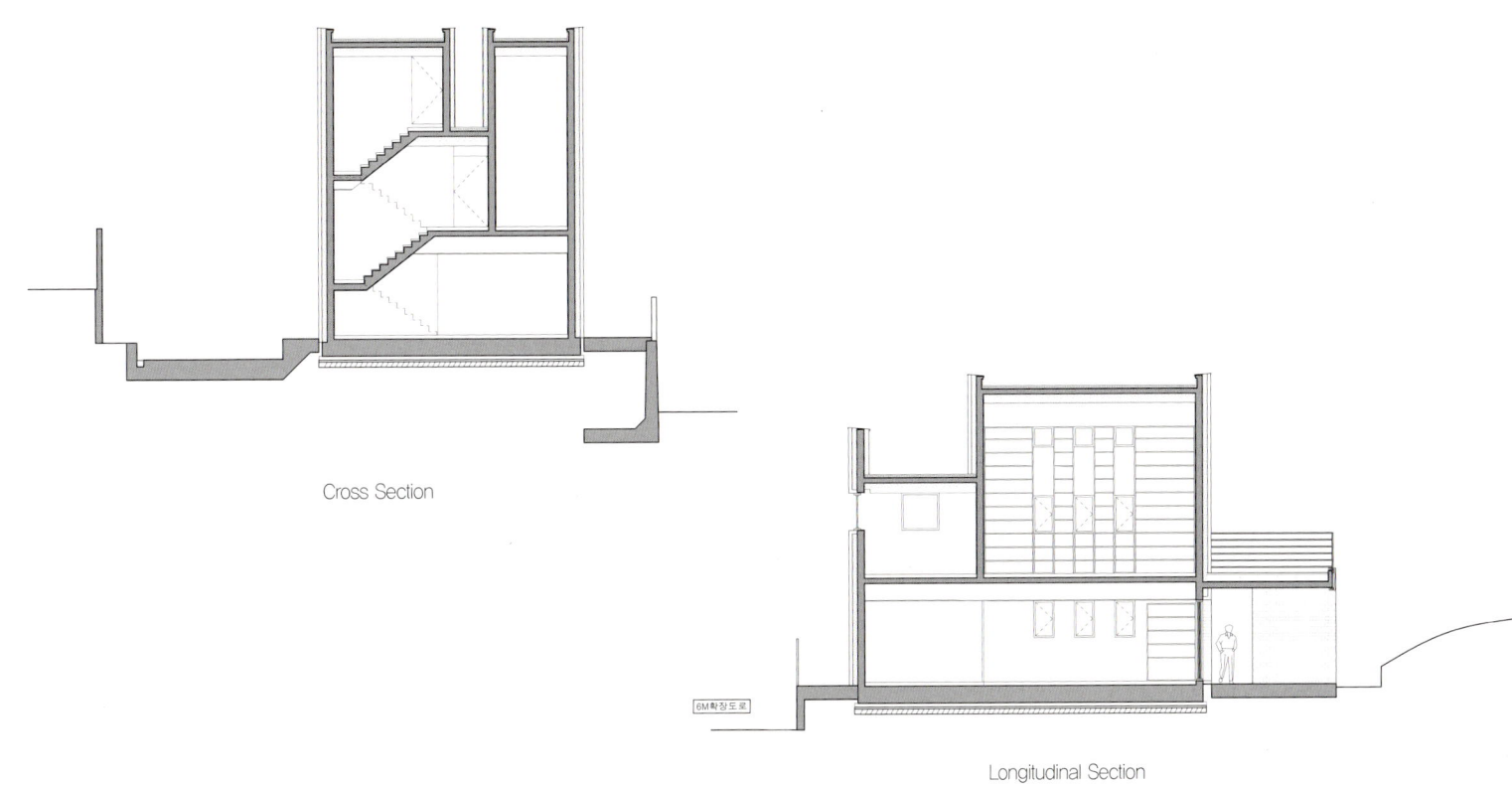

1. PARKING LOT
2. UTILITY ROOM
3. WAREHOUSE
4. BOILER ROOM
5. FRONT ROOM
6. KITCHEN
7. LIVING ROOM
8. LOGGIA

1st Floor Plan

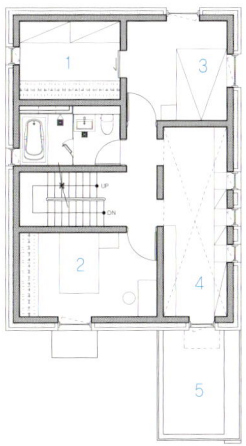
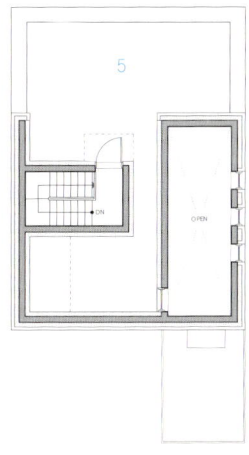

2nd Floor Plan Roof Plan

1. DRESS ROOM
2. NURSERY ROOM
3. MAIN ROOM
4. LIBRARY
5. TERRACE

Oreum Madang House Busan, Korea

JNPeople Architects

Architects JNPeople Architects / Sehwan Jang, Seonghee In **Location** Busan, Korea **Site Area** 418㎡ **Bldg. Area** 99.57㎡ **Gross Floor Area** 168.31㎡ **Bldg. Scale(Floor)** 2FL **Structure** Reinforced Concrete, Light Weight Wood Framing System **Max. Height** 8.31m **Exterior Finish** Stucco Flex **Photos Offer** JNPeople Architects

설계 제이앤피플 건축사사무소 / 장세환, 인성희 **위치** 부산 해운대구 중동 883-2외 1필지 **대지면적** 418㎡ **건축면적** 99.57㎡ **연면적** 168.31㎡ **규모** 지상2층 **구조** 철근콘크리트, 경량목구조 **최고높이** 8.31m **외부마감** 스터코플렉스 **사진제공** 제이앤피플 건축사사무소

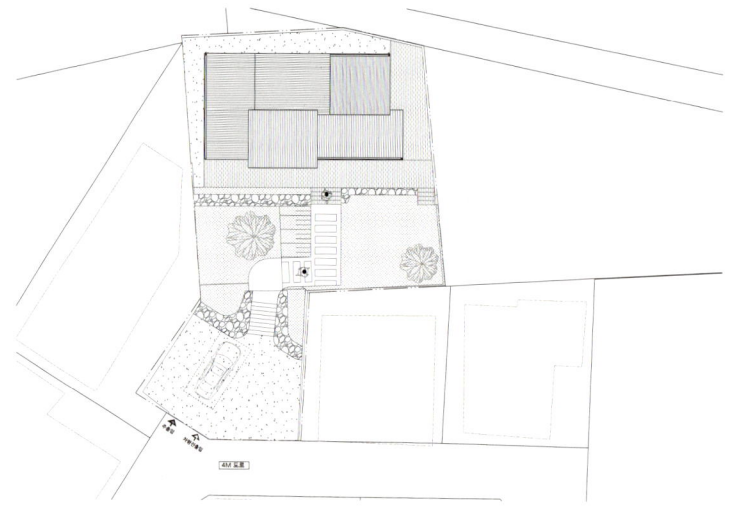

Site Plan

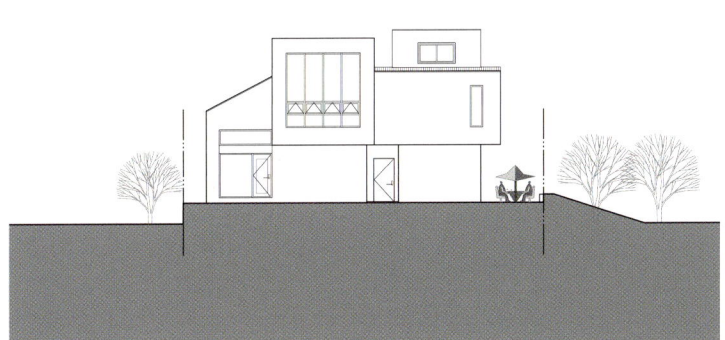

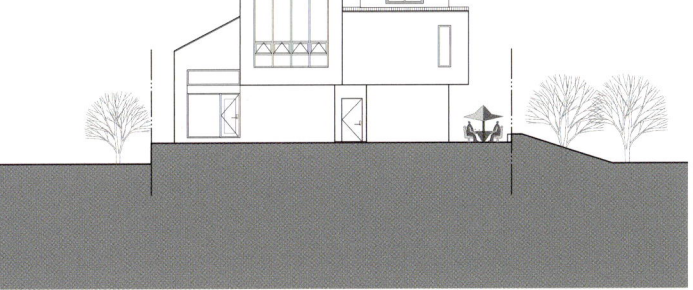
Front Elevation

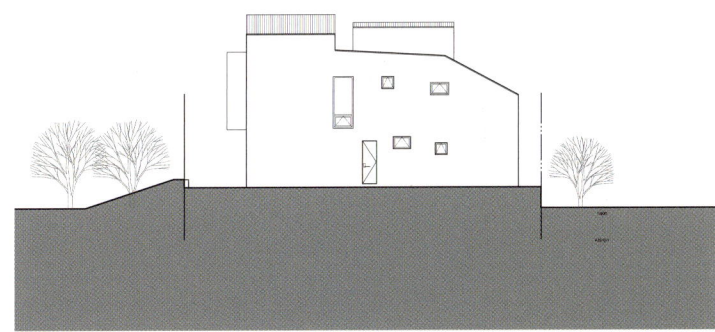
Rear Elevation

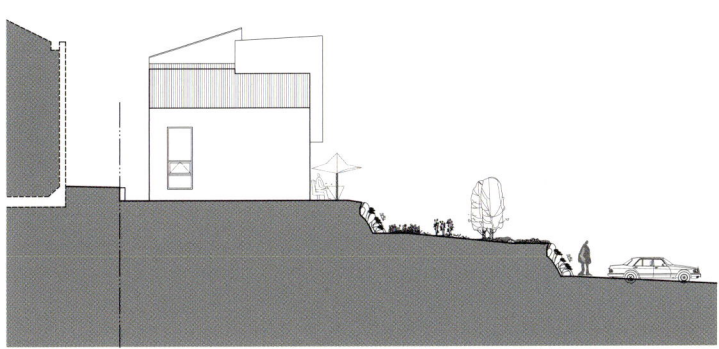
Left Elevation

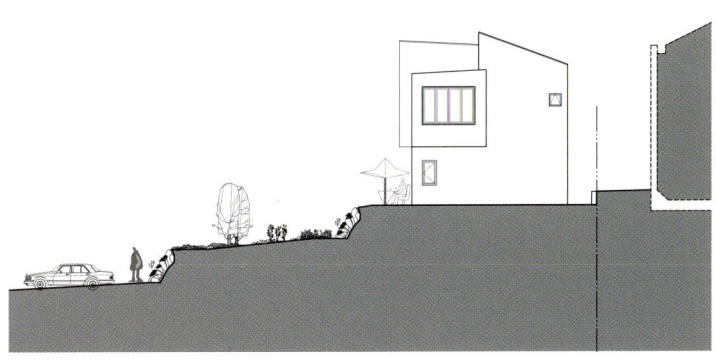
Right Elevation

'Oreum Madang House', a "space where differences coexist," is a detached house for a family including an elementary school student and three preschoolers in the Middle East of Haeundae-gu, Busan. A detached house is premised on a space composition that reflects the individuality of a family and a family rather than the generalized asset value like an apartment. Therefore, we need to objectively look into the life of the family through dialogue with the architect and restore and create the lost individuality one by one.

On the first floor, there is a family library for studying with three children. In consideration of the hilltop view, the living room and master bedroom are partitioned on the second floor, and the family room on the first floor has an interior window to look down on the family room so that communication between family members is possible. In the future, when children grow up and go home, the first floor can be used as a guest house. The space used by each family member is also reflected in the external form to reveal its individuality. Although it is not a large area of about 60 pyeong, it is designed to have a space where the individuality of all family members is respected and a community can be formed.

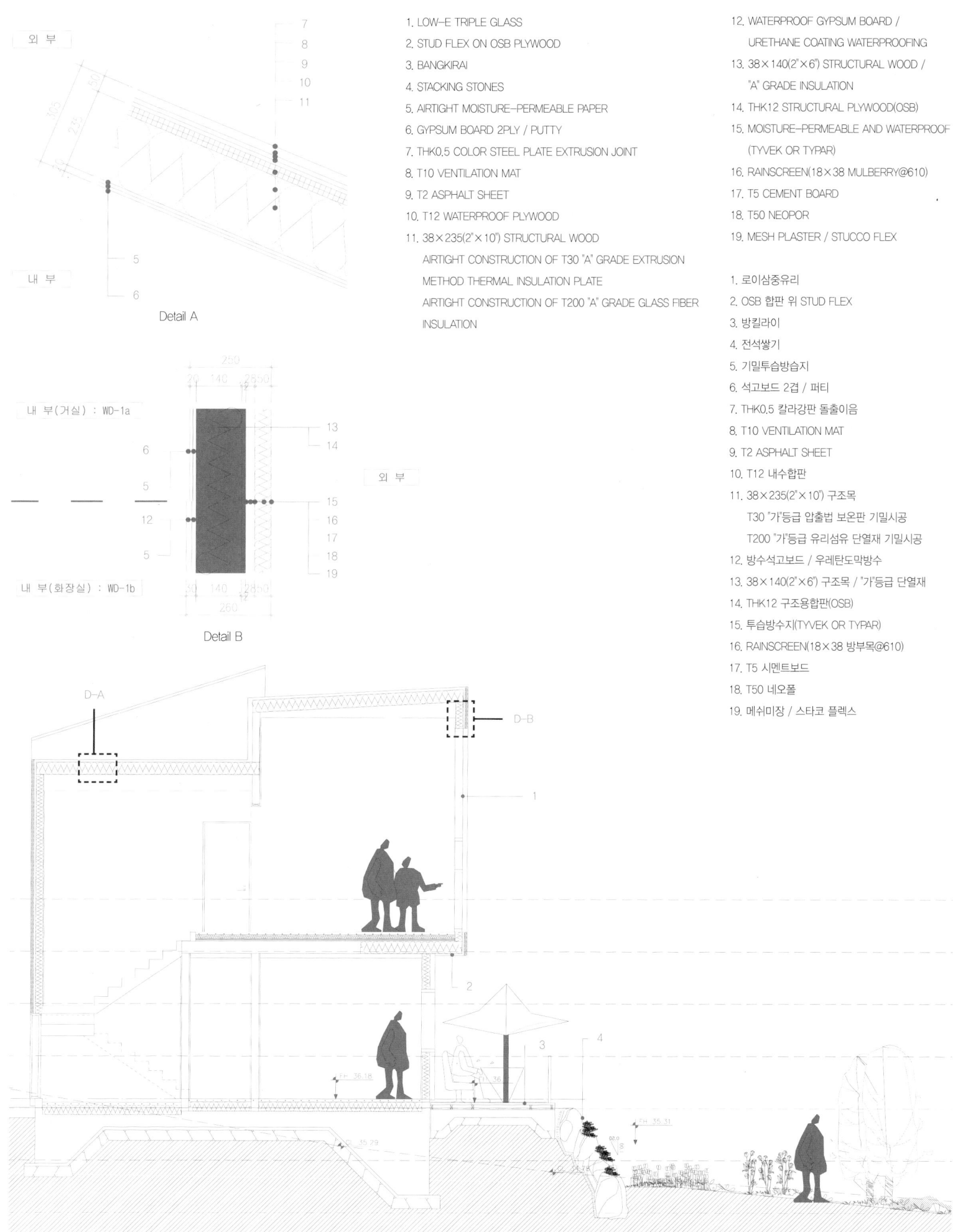

1. LOW-E TRIPLE GLASS
2. STUD FLEX ON OSB PLYWOOD
3. BANGKIRAI
4. STACKING STONES
5. AIRTIGHT MOISTURE-PERMEABLE PAPER
6. GYPSUM BOARD 2PLY / PUTTY
7. THK0.5 COLOR STEEL PLATE EXTRUSION JOINT
8. T10 VENTILATION MAT
9. T2 ASPHALT SHEET
10. T12 WATERPROOF PLYWOOD
11. 38×235(2"×10") STRUCTURAL WOOD
    AIRTIGHT CONSTRUCTION OF T30 "A" GRADE EXTRUSION METHOD THERMAL INSULATION PLATE
    AIRTIGHT CONSTRUCTION OF T200 "A" GRADE GLASS FIBER INSULATION
12. WATERPROOF GYPSUM BOARD / URETHANE COATING WATERPROOFING
13. 38×140(2"×6") STRUCTURAL WOOD / "A" GRADE INSULATION
14. THK12 STRUCTURAL PLYWOOD(OSB)
15. MOISTURE-PERMEABLE AND WATERPROOF (TYVEK OR TYPAR)
16. RAINSCREEN(18×38 MULBERRY@610)
17. T5 CEMENT BOARD
18. T50 NEOPOR
19. MESH PLASTER / STUCCO FLEX

1. 로이삼중유리
2. OSB 합판 위 STUD FLEX
3. 방킬라이
4. 전석쌓기
5. 기밀투습방습지
6. 석고보드 2겹 / 퍼티
7. THK0.5 칼라강판 돌출이음
8. T10 VENTILATION MAT
9. T2 ASPHALT SHEET
10. T12 내수합판
11. 38×235(2"×10") 구조목
    T30 "가"등급 압출법 보온판 기밀시공
    T200 "가"등급 유리섬유 단열재 기밀시공
12. 방수석고보드 / 우레탄도막방수
13. 38×140(2"×6") 구조목 / "가"등급 단열재
14. THK12 구조용합판(OSB)
15. 투습방수지(TYVEK OR TYPAR)
16. RAINSCREEN(18×38 방부목@610)
17. T5 시멘트보드
18. T50 네오폴
19. 메쉬미장 / 스타코 플렉스

Detail A

Detail B

Exterior Wall Detail

오름마당집

"다름이 공존하는 공간"인 '오름마당집'은 부산 해운대구 중동에 초등학생과 미취학 아동 3명을 포함한 가족을 위한 단독주택이다. 단독주택은 아파트처럼 범용화된 자산가치 보다는 한 가정 및 가족구성원의 개성을 반영한 공간구성을 전제로 한다. 그러므로 건축가와 대화를 통해 가족의 삶을 객관적으로 들여다 보고 하나씩 하나씩 잃어버린 개성을 되찾고 만들어 가야한다.

1층에는 세 자녀와 공부를 위한 가족도서관을 배치하였다. 언덕 위 조망을 고려해 거실과 안방은 2층에 구획하였으며, 1층의 가족실은 가족실을 내려다 볼 수 있도록 내부창을 두어 가족간 소통이 가능하도록 하였다. 추후 자녀들이 성장해 출가를 할 경우 1층은 민박용도로 사용이 가능하도록 하였다. 각각의 가족 구성원이 사용하는 공간은 외부형태로도 반영되어 그 개성이 드러나도록 하였다. 60평 내외의 넓지 않은 면적이지만 가족구성원 모두의 개성이 존중되며 커뮤니티가 형성될 수 있는 공간을 가질 수 있도록 구성하였다.

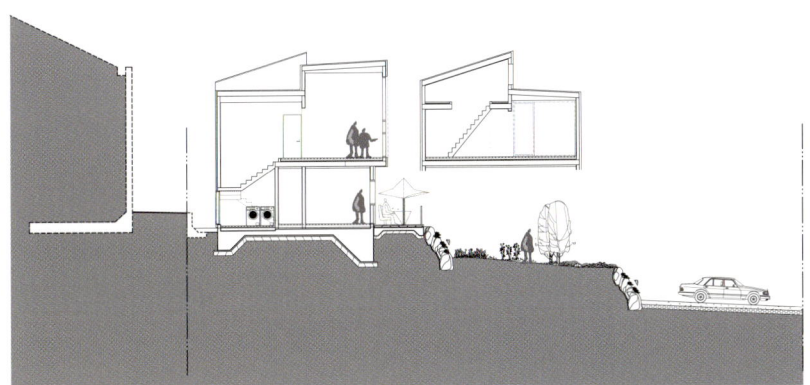

Longitudinal Section

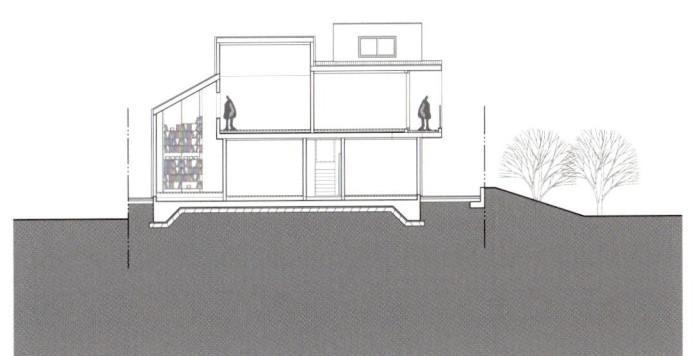

Cross Section

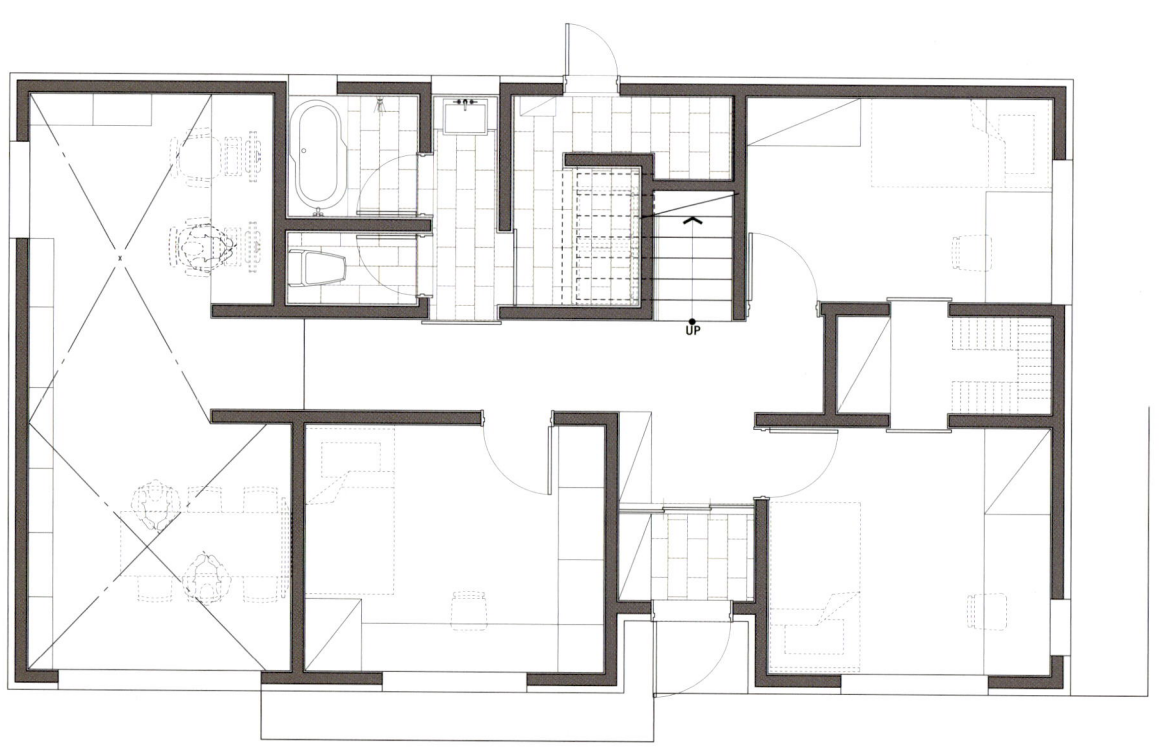

1st Floor Plan

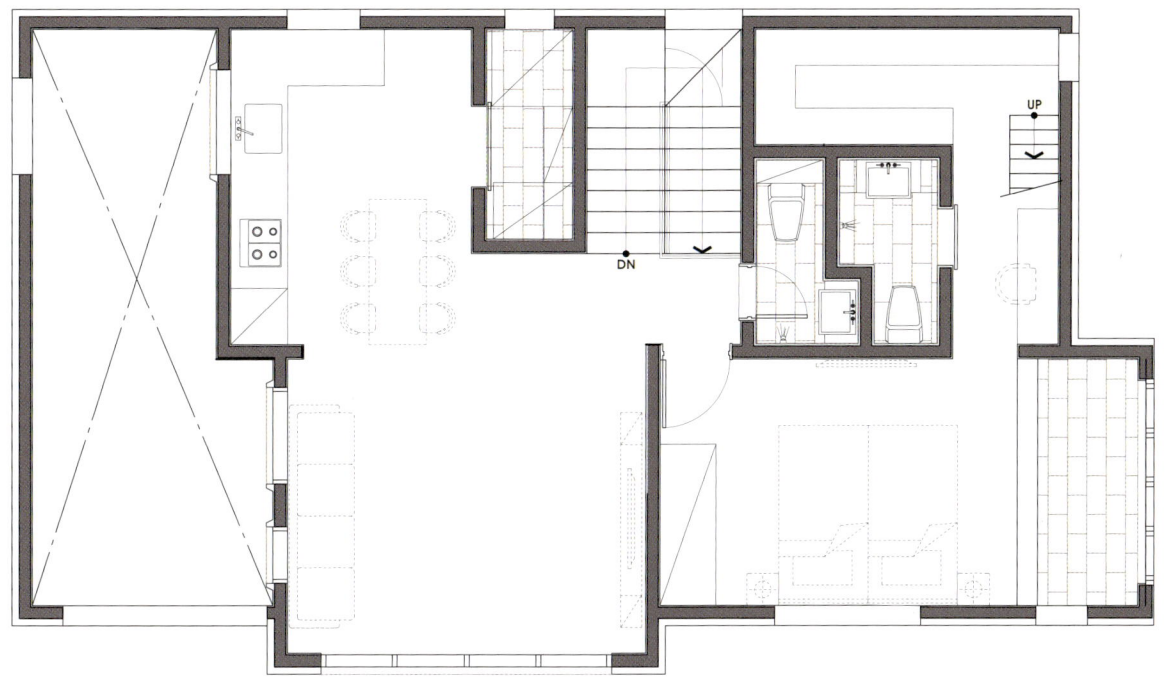

2nd Floor Plan

Cheongju Empty & Full House Cheongju, Korea

RICHUE Architecture

Architects RICHUE Architecture / Mansik Hong **Project Team** Youna Kim **Location** Cheongju, Korea **Site Area** 612㎡ **Bldg. Area** 112.07㎡ **Gross Floor Area** 184.76㎡ **Construture** WE · BUILD CONSTRUCTION **Photographer** Yongsun Kim, Jeongjung Kim

설계 ㈜리슈건축사사무소 / 홍만식 **설계참여자** 김유나 **위치** 청주시 흥덕구 월탄리 **대지면적** 612㎡ **건축면적** 112.07㎡ **연면적** 184.76㎡ **시공** 위빌종합건설 **사진** 김용순, 김정중

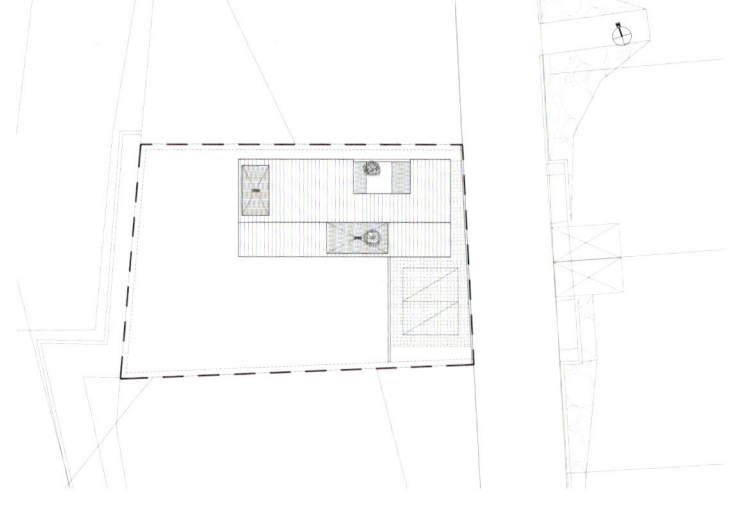

Site Plan

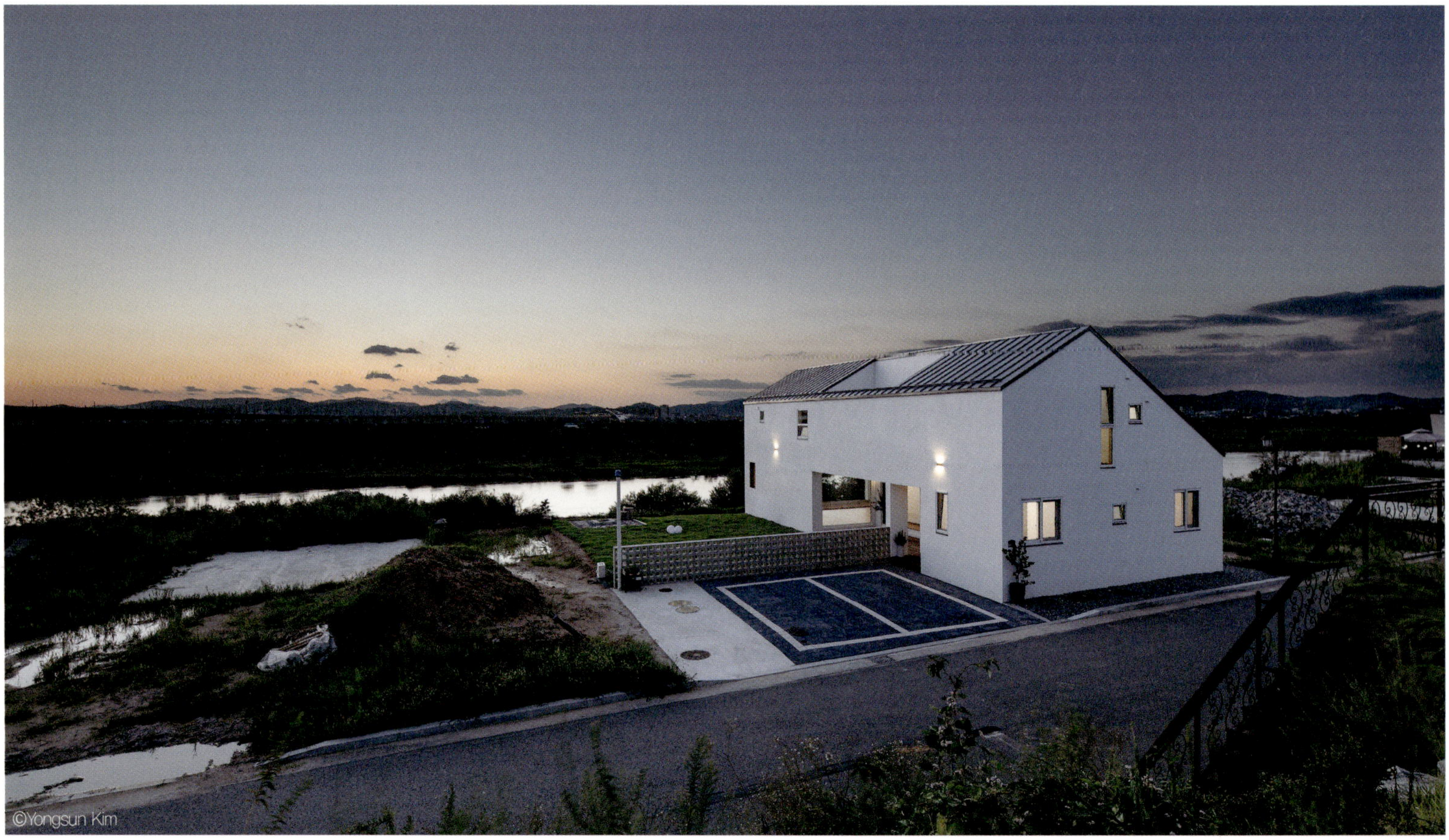

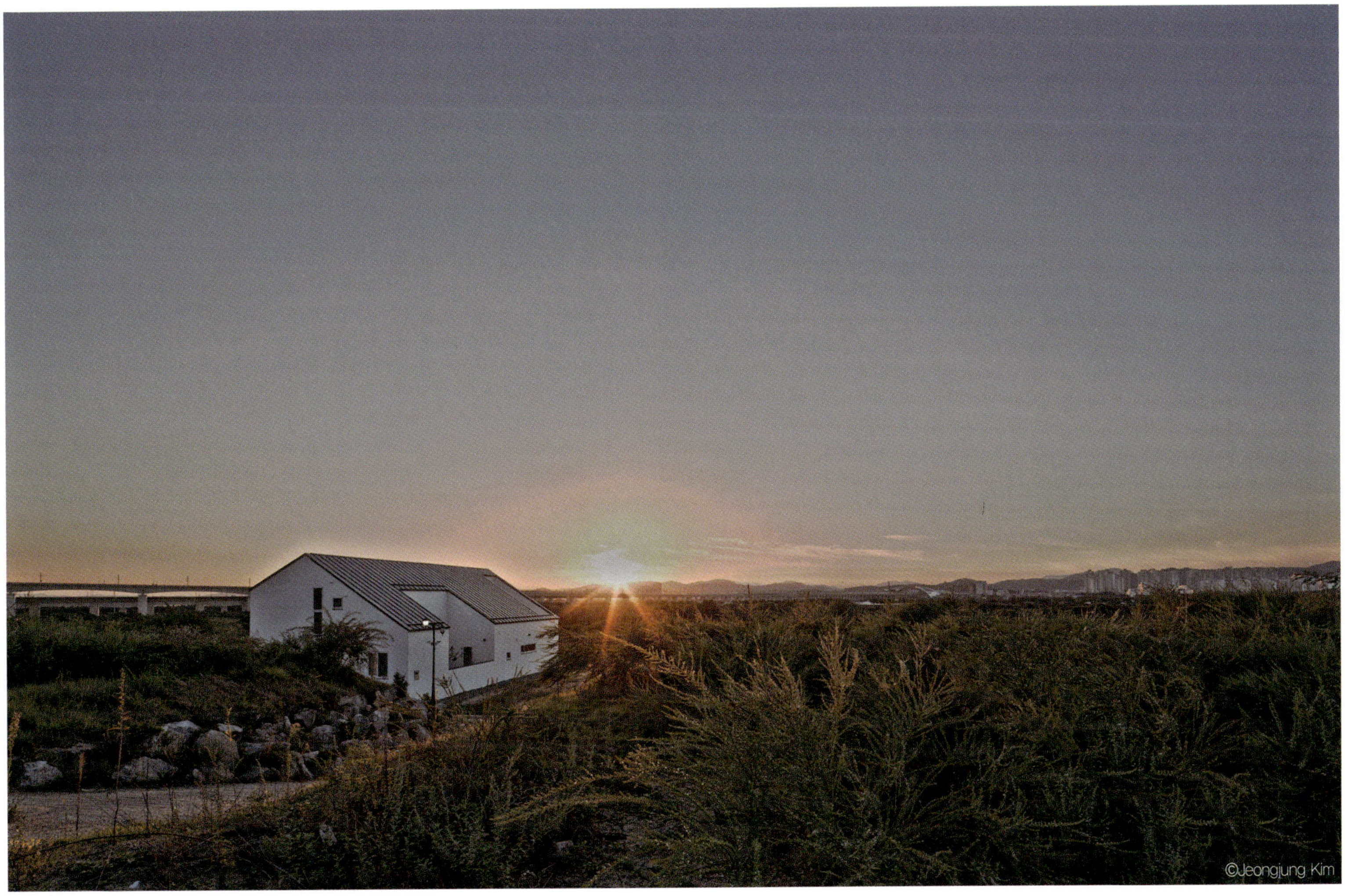

©Jeongjung Kim

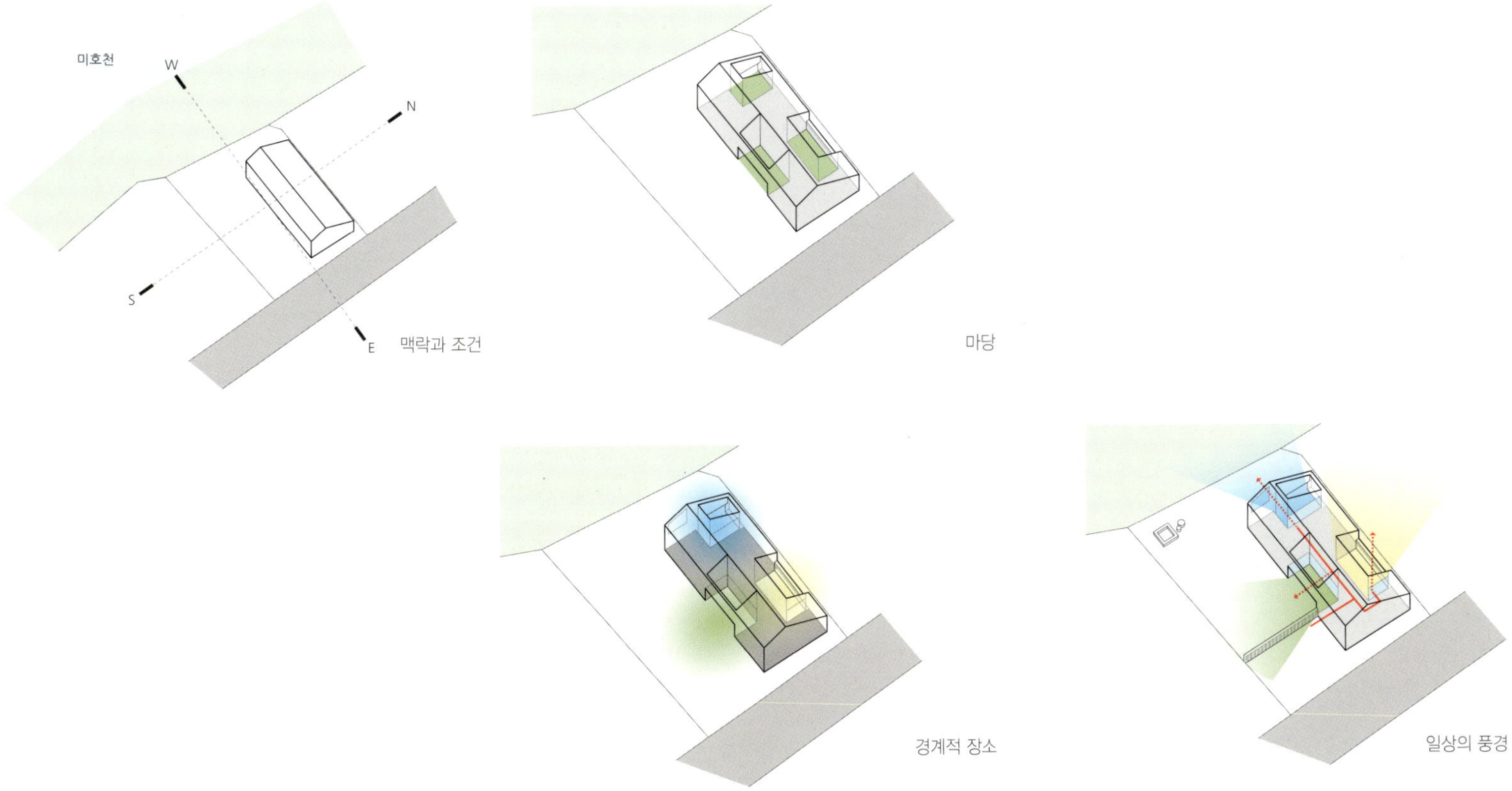

미호천 맥락과 조건

마당

경계적 장소

일상의 풍경

Diagram

81

Context and conditions
It is a land of 185pyeong in a housing complex with Mihocheon Stream near Cheongju to the west. It is a rectangular site that has an access road to the east and is slightly longer from east to west. The couple with their parents and two children always wanted their home to enjoy the scenery of Mihocheon Stream. Although the land was large, the cost was not sufficient, so we wanted to build a wooden house of about 40pyeong. I wanted it to be a living room and kitchen shared by everyone, even though they were separated from their parents. I wanted it to be a simple and rich house that matches the Mihocheon Stream and the house.

Form empty to contain
How can we organize the relationship between the vast land and the vast Mihocheon Stream with one small building? The plan started with a question. With the outer yard to the south, a simple mass that runs east to west in the west direction of Mihocheon Stream was planned in such a way that the surrounding landscape could be experienced as a distant view, a close view. I tried to change the way of enjoying the ever-changing scenery of nature like a picture frame through the windows and doors, just like the tea scenery of a hanok. We planned in a way that the simple mass was emptied of the west side of the Miho River, the south side of the outer yard, and the north side of the grandmother's room, and that it became a yard connected with threads and related it to the surrounding landscape.

The empty part to the south and west is the yard where the living room and kitchen are connected, creating a relationship with the landscape by placing the wall on the second floor to outline the mass. To the north is the yard where the grandmother's room and the utility room are connected, and a fence wall is placed on the first floor to capture the surrounding landscape. In a simple form, the empty yard works at the border with the surrounding landscape, becoming a device that creates a rich daily life in the flow of time.

The bordering place of the overlapping landscape
Surrounding landscapes overlap with the vacant yard and are experienced in various forms every moment. The empty yards work as a transitional space by intersecting each other rather than delimiting the inside and outside. In the western yard, you can see Mihocheon at a glance from the living room and kitchen. If it is opened using a folding door, the landscape of the outside and the living room and kitchen inside are experienced as one place. The southern yard is lit by sunlight, overlapping the large external yard, and looking at the central view. In addition to this, a single rhizome creates a rhizome, and both the rhizome and the middle rhizome work at the same time. The northern yard is connected with the grandmother's room, creating a close-up view. In addition, the near scene experiences the visual effect of overlapping the distant view beyond the wall. These vacant yards are becoming boundary places that create rich spatial experiences in everyday life while interacting with the landscape.

Interaction between daily life and landscape
Today's houses are demanding to be able to accommodate the various demands of life. It seems that he is always showing his desire to escape from everyday life even in his daily life. How can everyday life coexist? We wanted to create an event to escape from the daily life by infiltrating the landscape into the everyday space and interacting with the landscape. The daily spaces that interact with these landscapes are the living room and kitchen on the first floor, the grandmother's room on the first floor, the master bedroom and the bathroom with a view of the Mihocheon on the second floor, and the hallway on the first and second floors. These everyday spaces experience the coexistence of everyday life that is experienced as a new space depending on the penetration of light, the sunset scenery, the change of beech trees, and the change of weather even during the day. Here, the house is not a space filled with simple everyday functions, but a place where daily life and non-daily life coexist.

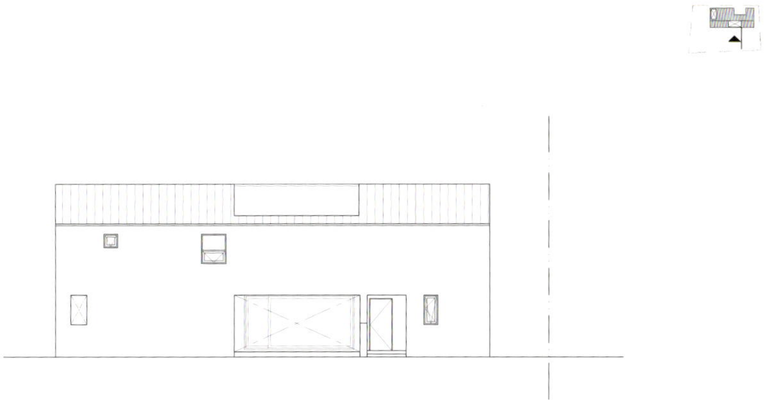

Elevation I

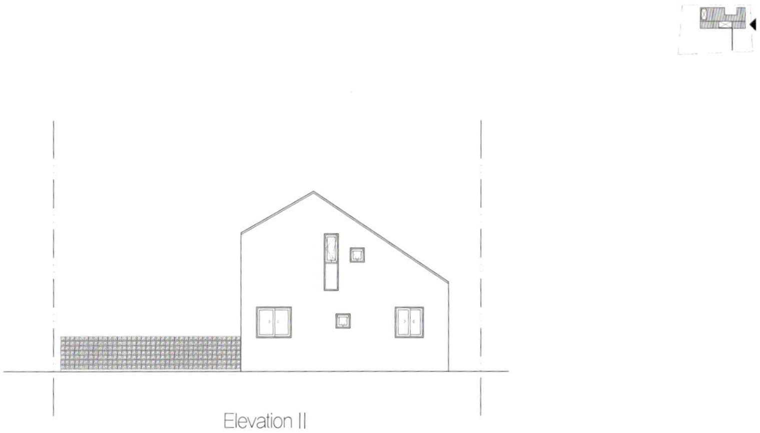

Elevation II

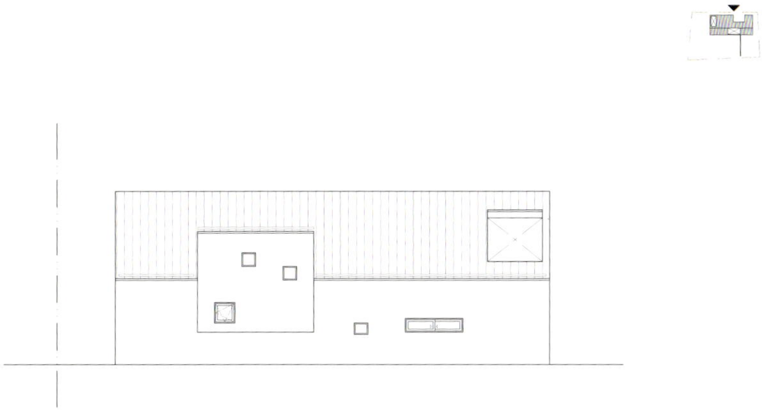

Elevation III

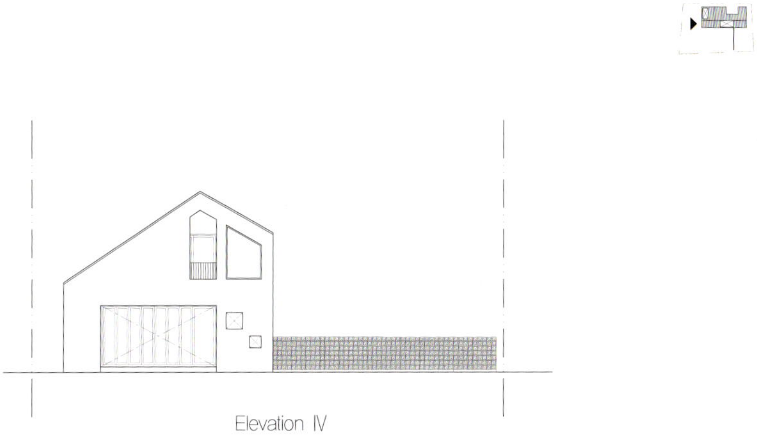

Elevation IV

©Jeongjung Kim

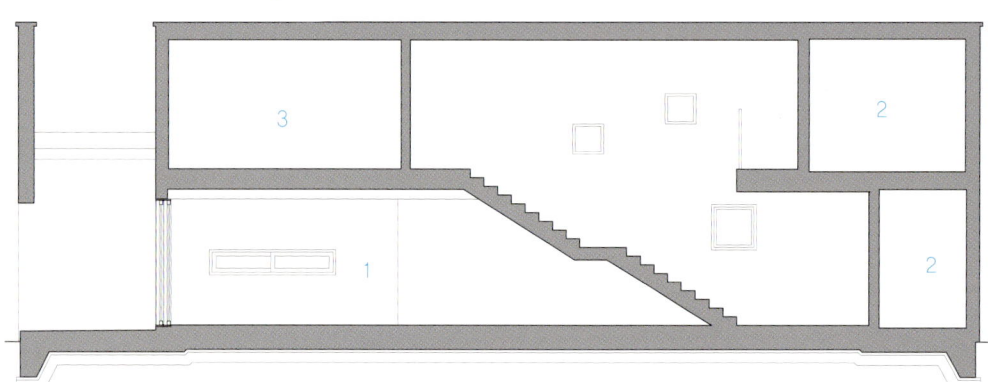

Section A

1. KITCHEN & LIVING ROOM
2. BATHROOM
3. ROOM
4. CORRIDOR

Section B

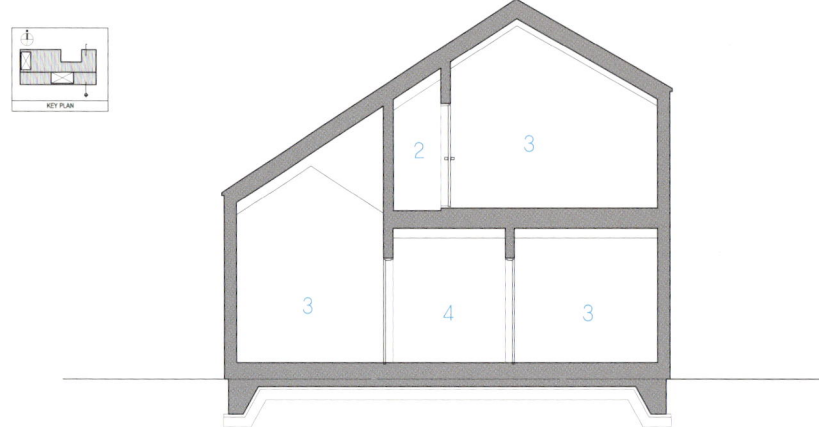

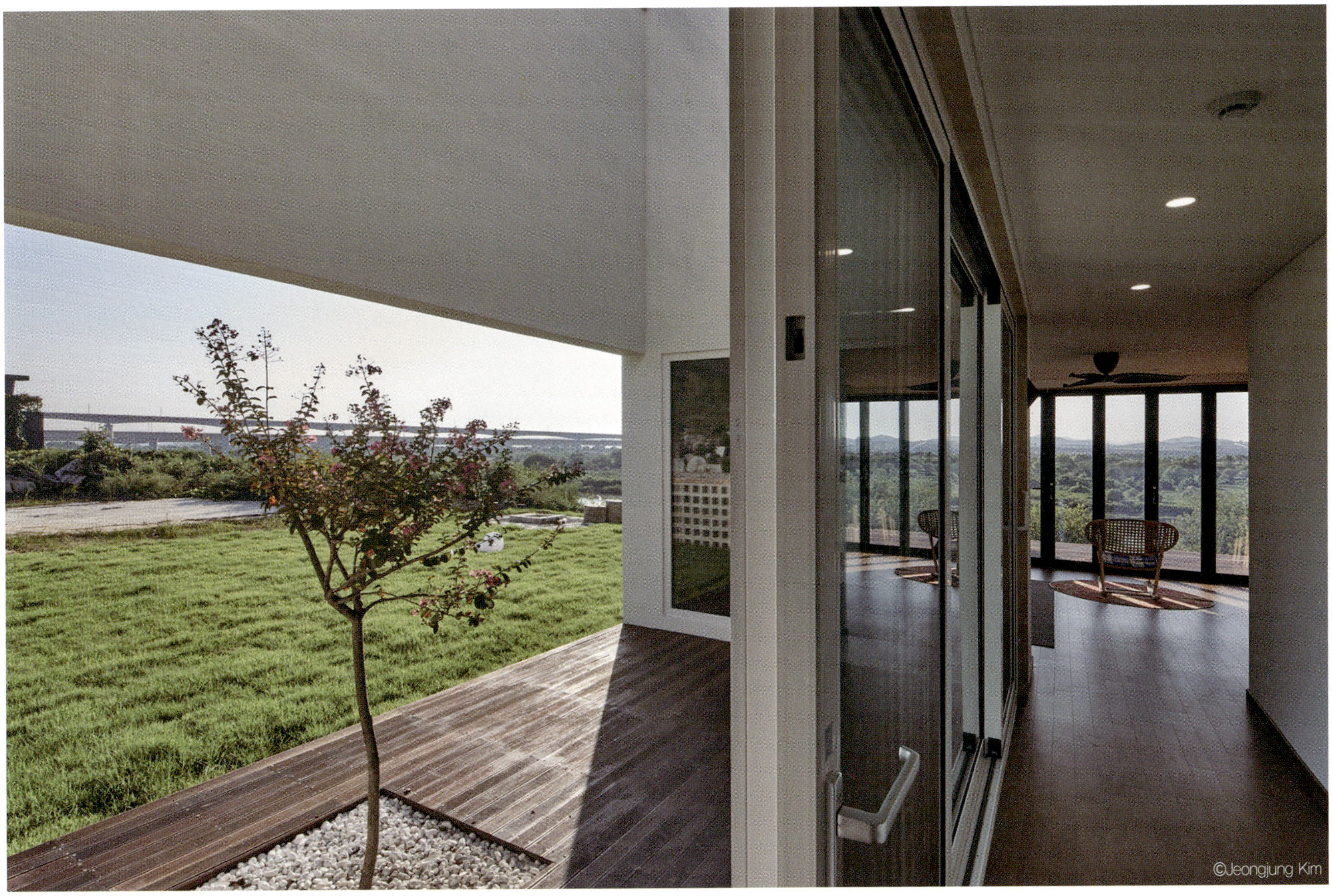

85

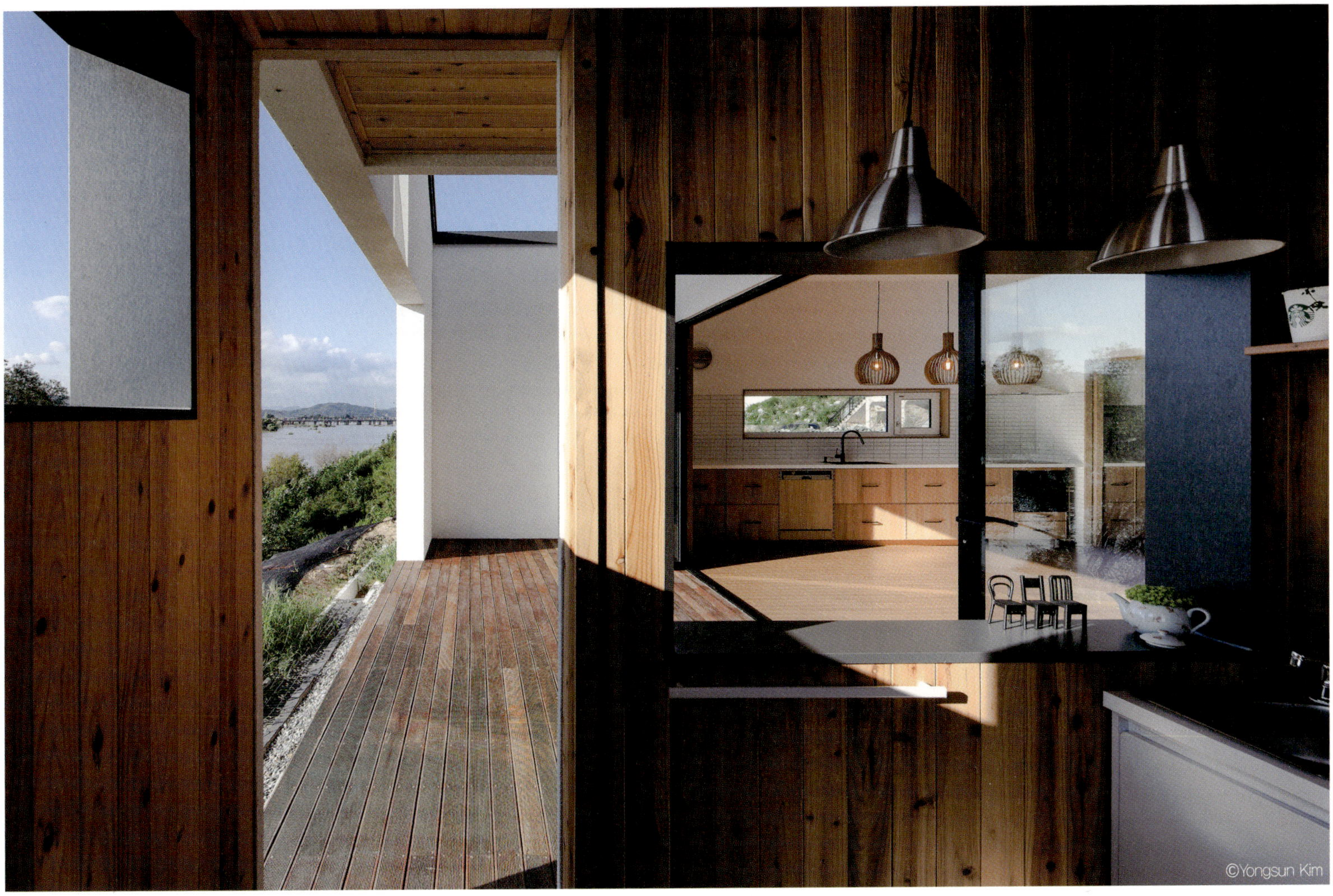

©Yongsun Kim

1. T0.5 GALVANIZED STEEL SHEET EXTRUSION JOINT
2. DELTA MEMBRANE
3. T2 SHEET WATERPROOF
4. T11.1 STRUCTURAL PLYWOOD
5. STRUCTURAL WOOD 38×68 @600
6. MOISTURE-PROOF PAPER FOR WARM ROOF (TYVEK SUPRO)
7. STRUCTURAL WOOD 38×235 @600 T235 GLASS WOOL R30(A GRADE)
8. T9.5 GYPSUM BOARD 2P
9. APP' WALLPAPER
10. T50 BEAD METHOD THERMAL INSULATION PLATE(A GRADE)
11. STUCCOFLEX
12. T140 GLASS WOOL R19(C GRADE)
13. T12.5 GYPSUM BOARD
14. T9 FLAT IRON RAILING (H=1,200) / COMBINATION PAINT
15. APP' WOOD FLOORING
16. Ø20 HOT WATER PIPE / T60 CEMENT MORTAR
17. PE FILM
18. T40 BEAD METHOD THERMAL INSULATION PLATE(B GRADE)
19. T20 NOISE PREVENTION MATERIAL BETWEEN FLOORS
20. T18.3 STRUCTURAL PLYWOOD
21. STRUCTURAL WOOD 38×235 @300
22. T235 GLASS WOOL R30 (MULTI-GRADE)
23. APP' T30 WOOD
24. T65 PLAIN CONCRETE (WM #8-150×150)
25. T115 EXTRUSION METHOD THERMAL INSULATION PLATE (PROVISIONAL GRADE)
26. PENETRATING WATERPROOF
27. ABANDONED CONCRETE
28. 2 LAYERS OF T0.03 PE FILM
29. T200 RUBBLE COMPACTION
30. TRANSPARENT OIL STAIN 2 TIMES
31. T21 RED CEDAR DECK
32. 40×40×1.6T SQUARE PIPE
33. CONCRETE BLOCK(Q BLOCK) DECORATION STACKING
34. Ø20 HOT WATER PIPE / T50 CEMENT MORTAR

1. T0.5 아연도강 강판 돌출이음
2. DELTA MEMBRANE
3. T2 쉬트방수
4. T11.1 구조용합판
5. 구조목 38×68 @600
6. 웜루프용 방습지(TYVEK SUPRO)
7. 구조목 38×235 @600 T235 글라스울 R30(가등급)
8. T9.5 석고보드 2P
9. 지정벽지
10. T50 비드법보온판(가등급)
11. 스타코플렉스
12. T140 글라스울 R19(다등급)
13. T12.5 석고보드
14. T9 평철난간(H=1,200) / 조합페인트
15. 지정 강마루
16. Ø20 온수파이프 / T60 시멘트몰탈
17. PE 필름
18. T40 비드법보온판(나등급)
19. T20 층간소음방지재
20. T18.3 구조용합판
21. 구조목 38×235 @300
22. T235 글라스울 R30(다등급)
23. T30 지정목재
24. T65 무근콘크리트(WM #8-150×150)
25. T115 압출법보온판(가등급)
26. 침투성방수
27. 버림콘크리트
28. T0.03 PE필름 2겹
29. T200 잡석다짐
30. 투명오일스테인 2회
31. T21 적삼목데크
32. 40×40×1.6T 각파이프
33. 콘크리트블럭(큐블럭) 치장쌓기
34. Ø20 온수파이프 / T50 시멘트몰탈

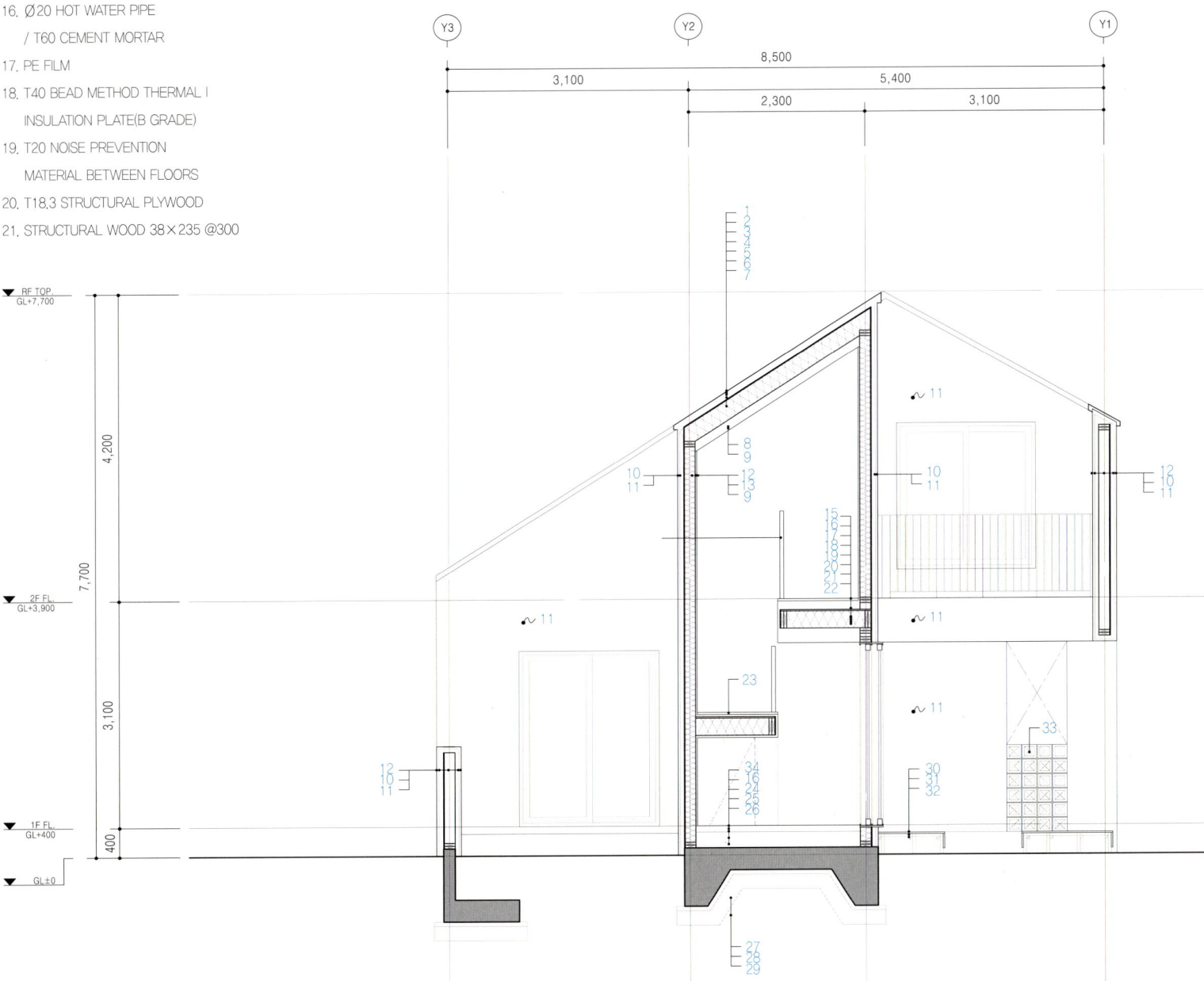

Stair Detail

청주 비담집(비우고 담은 집)

맥락과 조건
청주 근교의 미호천을 서쪽으로 둔 주택단지 내 185평의 대지이다. 동쪽에 진입도로가 있으면서 동서로 조금 긴 직사각형의 대지이다. 부모님을 모시면서 두 자녀를 둔 부부는 미호천 풍경을 집이 항상 누리기를 원했다. 넓은 땅이지만 비용이 넉넉히 못해 40평 남짓한 목조주택을 짓고자 했다. 부모님과 분리되면서도 다 같이 공유하는 거실과 주방이기를 원했다. 미호천 풍경과 집이 어울리는 단순하면서 풍부한 집이기를 원했다.

담기 위해 비운 형태
우리는 작은 건축물 하나로 넓은 대지와 광활한 미호천을 어떻게 관계 조직할 수 있을까? 하는 질문으로 계획을 시작했다. 남쪽으로 외부마당을 두고 서쪽 미호천 방향으로 동서로 긴 단순한 매스를 계획하면서 주변풍경을 원경 중경 근경으로 경험 할 수 있는 방식으로 계획해 갔다. 한옥의 차경처럼 창과 문을 통해 시시각각 변하는 자연의 경치를 액자처럼 누리는 방식을 변용하고자 했다. 우리는 단순한 매스를 미호천인 서쪽, 외부마당인 남쪽, 할머니 방 쪽인 북쪽을 비우고, 실들과 연계된 마당이 되면서 주변 풍경과 관계 짓는 방식으로 계획을 해 갔다.

남쪽과 서쪽의 비워진 부분은 거실과 주방이 연계되는 마당으로 2층 벽을 두어 매스의 윤곽을 만들면서 풍경과 관계를 만들고 있다. 북쪽은 할머니방과 다용도실이 연계되는 마당으로 1층에 담장 벽을 두면서 주변 풍경을 담아 내고 있다. 단순한 형태에서 비워진 마당은 주변 풍경과의 경계에서 작동하여 시간의 흐름 속에서 풍부한 일상을 만드는 장치가 되고 있다.

중첩된 풍경의 경계적 장소
주변 풍경들은 비워진 마당과 중첩되면서 시시각각 다양한 모습으로 경험되어진다. 비워진 마당들은 내·외부를 경계 짓기 보다는 서로 교차 시키면서 전이적 공간의 성격으로 작동한다. 서쪽 마당은 거실과 주방에서 미호천 원경을 한눈에 바라 볼 수 있다. 폴딩 도어를 사용하여 열어 둘 경우 외부인 풍경과 내부인 거실과 주방은 한 장소로 경험된다. 남쪽 마당은 채광을 받으면서 넓은 외부 마당을 중첩시키면서 중경으로 바라보게 된다. 여기에 더해 한그루의 베롱나무는 근경을 만들면서 근경과 중경이 동시에 작동한다. 북쪽 마당은 할머니 방과 연계되면서 근경을 만들어 내고 있다. 또한 근경은 담장 너머 원경이 중첩되는 시각적 효과를 경험을 하게 된다. 이처럼 비워진 마당들은 풍경과 상호작용하면서 일상의 삶에 풍부한 공간적 경험을 만드는 경계적 장소가 되고 있다.

일상과 풍경의 상호작용
요즘 집은 다양한 삶의 요구를 담을 수 있게끔 요구를 하신다. 일상을 살면서도 항상 탈 일상의 욕구를 드러내고 있는 듯 하다. 일상 속 탈 일상이 어떻게 공존할 수 있을까? 우리는 일상의 공간 속에 풍경을 침투시켜 풍경과 상호작용하는 방식으로 탈 일상의 이벤트를 만들고자 했다. 이러한 풍경과 상호 작용하는 일상의 공간은 1층의 거실과 주방, 1층의 할머니방, 2층의 안방과 미호천이 보이는 욕실, 1, 2층의 복도이다. 이들 일상의 공간들은 하루 중에도 빛의 침투나 석양 풍경, 베롱나무의 변화, 날씨변화 등에 따라 새로운 공간으로 경험되는 탈 일상의 공존을 경험하게 된다. 여기서 집은 단순한 일상적 기능으로 채워진 공간이 아니라 일상과 풍경이 상호작용하여 만들어내는 일상과 탈 일상이 공존하는 장소가 되는 것이다.

1. ENTRANCE
2. ROOM
3. BATHROOM
4. KITCHEN & LIVING ROOM
5. UTILITY ROOM
6. WAREHOUSE
7. TERRACE
8. COURTYARD
9. YARD
10. LIBRARY
11. DRESS RIOOM

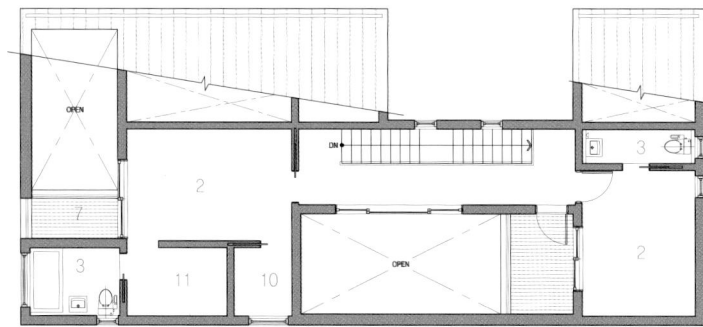

2nd Floor Plan

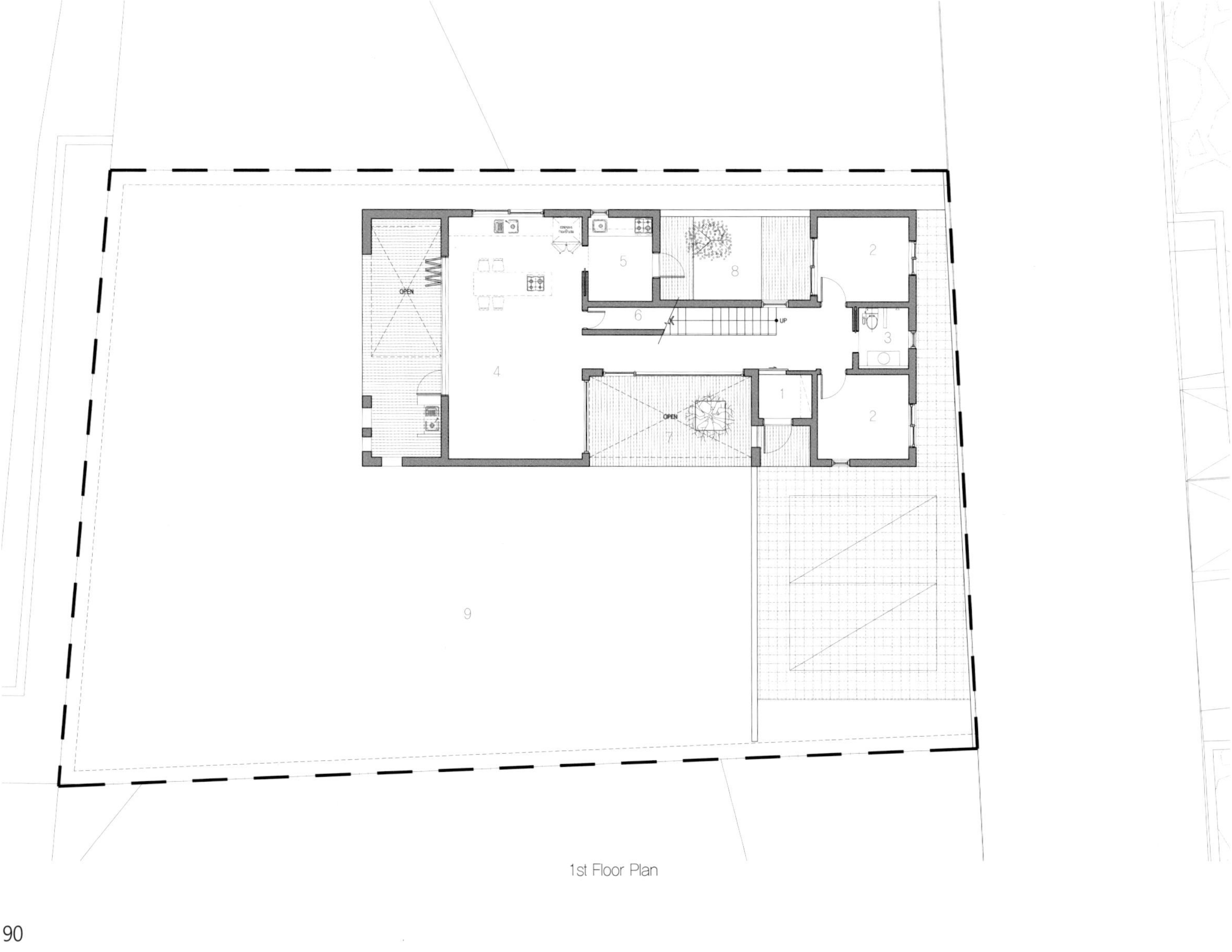

1st Floor Plan

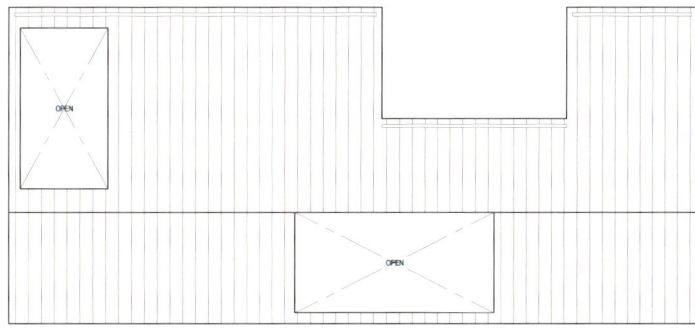

Roof Plan

House_B382 Yangpyeong, Korea

Moku Architects

Architects Moku Architects / Junghoon Mok , Jungho Mok **Location** Yangpyeong, Korea **Site Area** 391㎡ **Bldg. Area** 112.55㎡ **Gross Floor Area** 149.93㎡ **Bldg. Scale(Floor)** 2FL **Structure** Reinforced Concrete **Exterior Finish** Stuccoflex, Cedar, Corrugated Steel Plate **Photographer** Studio Greysome / Junhwan Kim

설계 ㈜모쿠아키텍츠 / 목정훈, 목정호 **위치** 경기도 양평군 서종면 수능리 일반 382-3 **대지면적** 391㎡ **건축면적** 112.55㎡ **연면적** 149.93㎡ **규모** 지상2층 **구조** 철근콘크리트 **외부마감** 스타코 플렉스, 삼나무, 골강판 **사진** 스튜디오 그레이썸 / 김준환

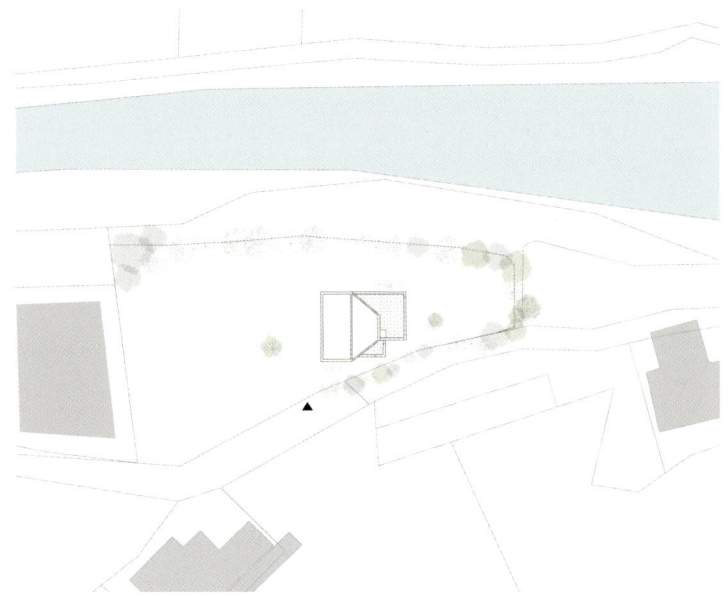

Site Plan

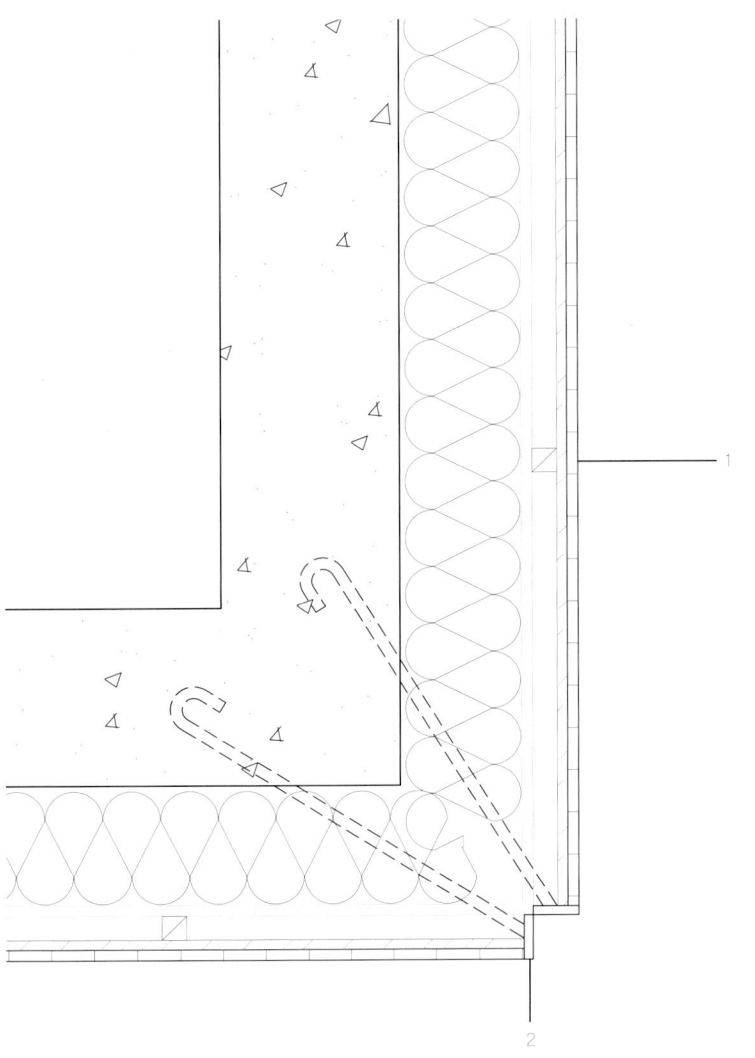

1. THK200 콘크리트 벽체
   THK135 압출법보온판 가등급
   12MM 방수몰탈
   27×27 각재
   11MM OSB 합판
   12MM 삼나무 사이딩
2. 외부 디자인 난간 지지용 인입 앵커
   THK10 외부 디자인 철제 난간

1. THK200 CONCRETE WALL
   THK135 EXTRUSION METHOD THERMAL INSULATION PLATE A GRADE
   12MM WATERPROOF MORTAR
   27×27 SQUARE WOOD
   11MM OSB PLYWOOD
   12MM CEDAR SIDING
2. INLET ANCHOR FOR SUPPORTING EXTERNAL DESIGN HANDRAIL
   THK10 EXTERIOR DESIGN IRON RAILING

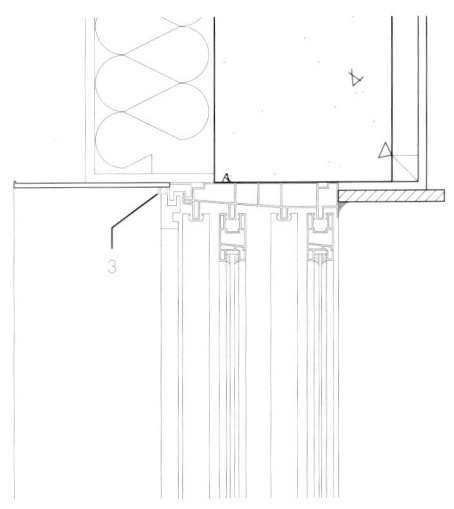

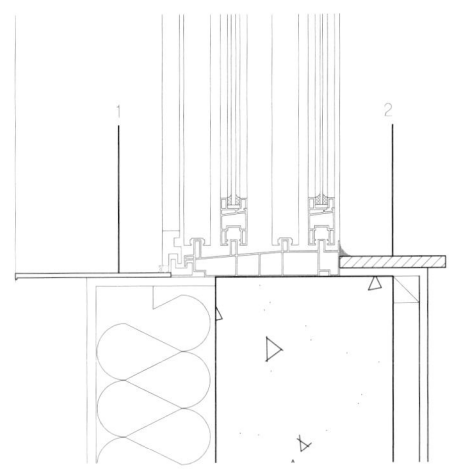

1. THK135 압출법보온판 가등급
   THK10 스타코 마감
   5MM 알루미늄
2. THK200 콘크리트 벽체
   9.5MM 석고보드
   18MM 합판
3. 실리콘 코킹(투명색)

1. THK135 EXTRUSION METHOD THERMAL INSULATION PLATE A GRADE
   THK10 STUCCO FINISH
   5MM ALUMINUM
2. THK200 CONCRETE WALL
   9.5MM GYPSUM BOARD
   18MM PLYWOOD
3. SILICONE CAULKING (TRANSPARENT)

This building is a country house for an elderly couple. Located in Seojong-myeon, Yangpyeong, Gyeonggi-do, it is relatively close to Seoul, but it is also a place where you can experience rural life while enjoying a rich natural environment.
The name of the house, B382, was inspired by the name of the asteroid B612, where the little prince lived. We had many drawing meetings and stories of life with the owner, who is preparing for a rural life with nature, while keeping the many memories and times that he has experienced while moving from city life to living in an apartment building. This is because the story that came to mind when I saw an elderly couple who were feeling the thrill of a new beginning in an unfamiliar place and even the fear of country life was the novel The Little Prince by Saint-Exupéry. The owner couple, starting with living in a detached house, moved through several moves, leaving behind the living life in the apartment that had been adapted to, and for what reason they wanted to return to living in a detached house again, and as a result, we I came up with the idea of creating their own asteroid where a new story begins for this old couple.
To the north of the site, because it includes a river site, there are legal restrictions that cannot allow construction to take place as much as the visible space. It was not easy to explore the relationship between the architectural scale and nature. In order to solve these environmental constraints, we arrange smaller masses as we go southeast based on the east and west axes to prepare a garden that naturally connects from the entrance to the site through the front door, and at the same time, we have high windows and Void space was provided to secure colorful lighting and scenery.

This building, with a total floor area of 150㎡, has a flat surface consisting of a combination of three rectangular masses lined up according to the shape of the site. At the same time, a complementary relationship between the outside and the inside was formed. And the crown-shaped roof with various angles tying them together seeps into the surrounding background as if the mountain ridges are connected, and surrounds the two veranda spaces with different directions, allowing the scenery of the mountains, gardens, and rivers to be fully embraced.
As a result of the internal space program, most of the living was planned on the first floor in consideration of the residential environment of the elderly couple. As a result, the first floor had a space full of building-to-land ratio of 112㎡, and the second floor was a small space of 38㎡ with a guest room and a small living room. It was planned.
On the first floor, there is an entrance hall with ample storage, a utility room, and an auxiliary kitchen. In addition, the living room with three wide windows with a high ceiling was designed considering the natural elements of the outside to be enjoyed in the interior space. The open feeling can be extended to the kitchen. In addition, the internal staircase including the toilet is placed in the center of the building and acts as a barrier separating the master bedroom, living room, entrance, kitchen, and guest space.

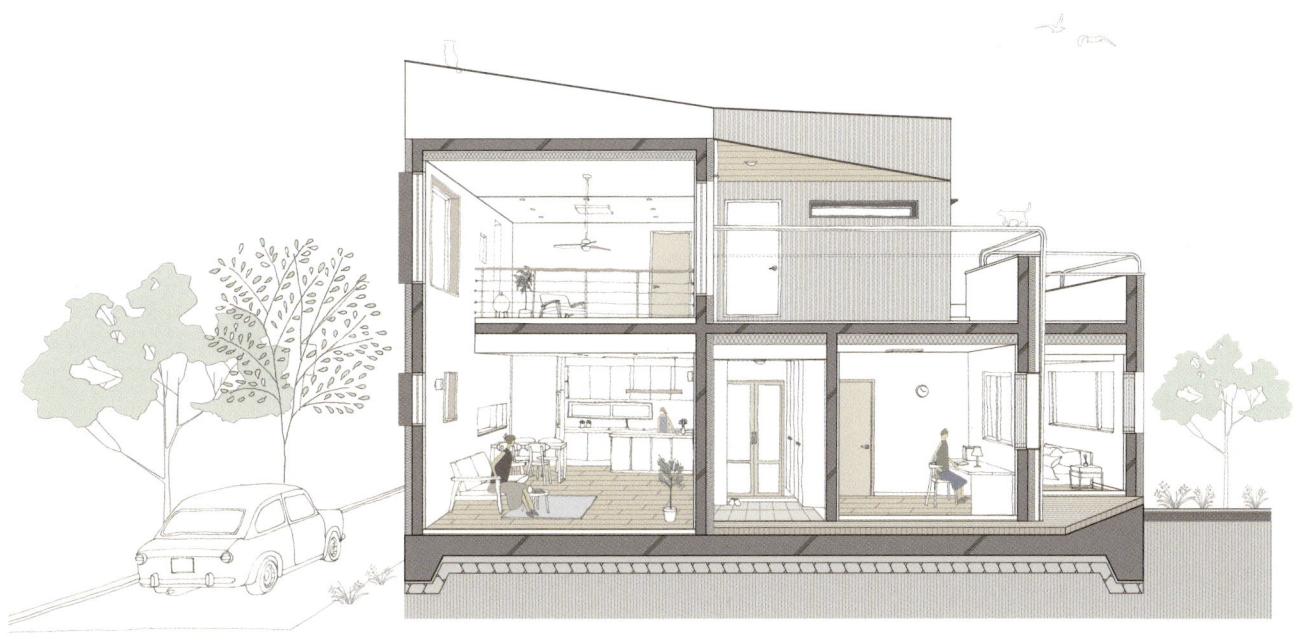
Section

하우스_ B382

이 건물은 노부부를 위한 전원주택이다. 경기도 양평의 서종면에 자리하며 비교적 서울과 근접하면서도 풍요로운 자연환경을 누리며 전원의 생활을 경험할 수 있는 곳에 위치해 있기 때문에 다양한 연령대의 거주민과 라이프스타일을 엿볼 수 있는 지역이기도 하다.

이 주택의 이름인 B382는 어린왕자가 살고 있던 B612라는 소행성의 이름에서 영감을 얻어 가져온 것이다. 도시생활에서 여러 번의 이사와 아파트 주거생활을 거치면서 겪었던 수많은 추억과 시간을 간직한 채 자연과 함께하는 전원생활을 준비하는 건축주와 많은 도면미팅과 삶의 이야기를 나누었다. 낯선 곳에서의 새로운 시작의 설레임과 전원생활에 대한 두려움까지 느끼고 있는 노부부를 보면서 문득 떠오르게 된 이야기가 바로 쌩텍쥐페리의 소설 어린왕자였기 때문이다. 건축주 부부는 단독주택의 주거를 시작으로 여러 번의 이사를 거치면서 적응되어진 아파트에서의 주거생활을 뒤로한 채 어떠한 이유로 또 다시 단독주택에서의 생활로 돌아가려 하는지 우리에게 깊은 고민을 하게 하였으며, 그 결과 우리는 이 노부부에게 새로운 이야기가 시작되는 그들만의 소행성을 만들어 드리고 싶다는 생각을 하게 되었다.

부지의 북쪽으로는 하천부지를 포함하고 있기 때문에 눈에 보이는 공간만큼 건축행위를 할 수 없는 법규상의 제약이 있었으며 남쪽으로는 청계산 기슭에 접해 있어서 남쪽채광에 대한 불리함을 가지고 있는 기존 환경의 특성상 새로운 건축적 스케일과 자연과의 관계를 모색하기란 쉽지 않았다. 우리는 이러한 환경적 제약을 해결하기 위해 동과 서의 축을 기준으로 동남쪽으로 향할수록 점점 작은 매스를 배치하여 부지의 입구에서부터 현관을 지나 자연스럽게 이어지는 정원을 마련함과 동시에 산자락의 풍경을 그대로 받아들이는 높은 창과 보이드 공간을 두어 다채로운 채광과 풍경을 확보하였다.

총바닥면적이 150㎡을 가지는 이 건축은 대지의 형태에 따라 늘어선 세 개의 직방형 매스의 조합으로 이루어진 평면을 가지며 이 세 개의 매스의 크기와 부피를 각각 다르게 설정함으로서 부지의 형태와 건물의 관계가 다양해짐과 동시에 외부와 내부의 상호보완적인 관계가 형성되었다. 그리고 그것들을 하나로 묶는 다양한 각도를 가지는 왕관 형태의 지붕은 산의 능선이 이어지는 듯 주변 배경속에 스며들며 서로 다른 방향을 가지는 두개의 베란다공간을 에워싸며 산과 정원, 그리고 하천의 풍경을 고스란히 받아들일 수 있게 하였다.

내부공간의 프로그램은 노부부의 주거환경을 고려하여 대부분의 생활을 1층에서 이루어지게 계획한 결과 1층은 112㎡로 건폐율 가득한 공간을 가지게 되었으며 2층은 게스트룸과 작은 거실을 가지는 38㎡의 아담한 공간으로 계획되었다.

1층에는 여유로운 수납을 가지는 현관과 다용도실, 보조주방이 있으며 남편을 위한 서재는 욕실과 드레스룸이 달린 침실과 더불어 정원과 마주하게 계획하였다. 또한 높은 천장과 함께하는 세 개의 넓직한 창을 가지는 거실은 외부의 자연적 요소를 내부공간에서 그대로 즐길 수 있도록 고려하여 설계한 결과 건축의 인공적 스케일이라기 보다는 자연의 스케일에 가까운 공간감을 선사하며 북측에 배치된 주방까지 개방감을 확장할 수 있었다. 또한 화장실을 포함하는 내부 계단은 건물 중앙에 배치하여 안방과 거실 및 현관, 주방 그리고 2층의 게스트 공간을 구분하는 경계로써 작용하며 2층의 복도 공간은 동쪽과 남쪽 두개의 옥상테라스를 이어주는 동선으로 계획되어 외부와 내부를 자유로이 순환할 수 있게 하였다.

97

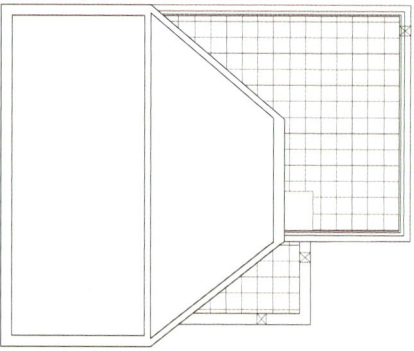

Roof Plan

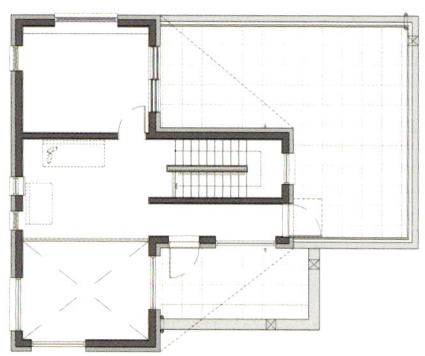

2nd Floor Plan

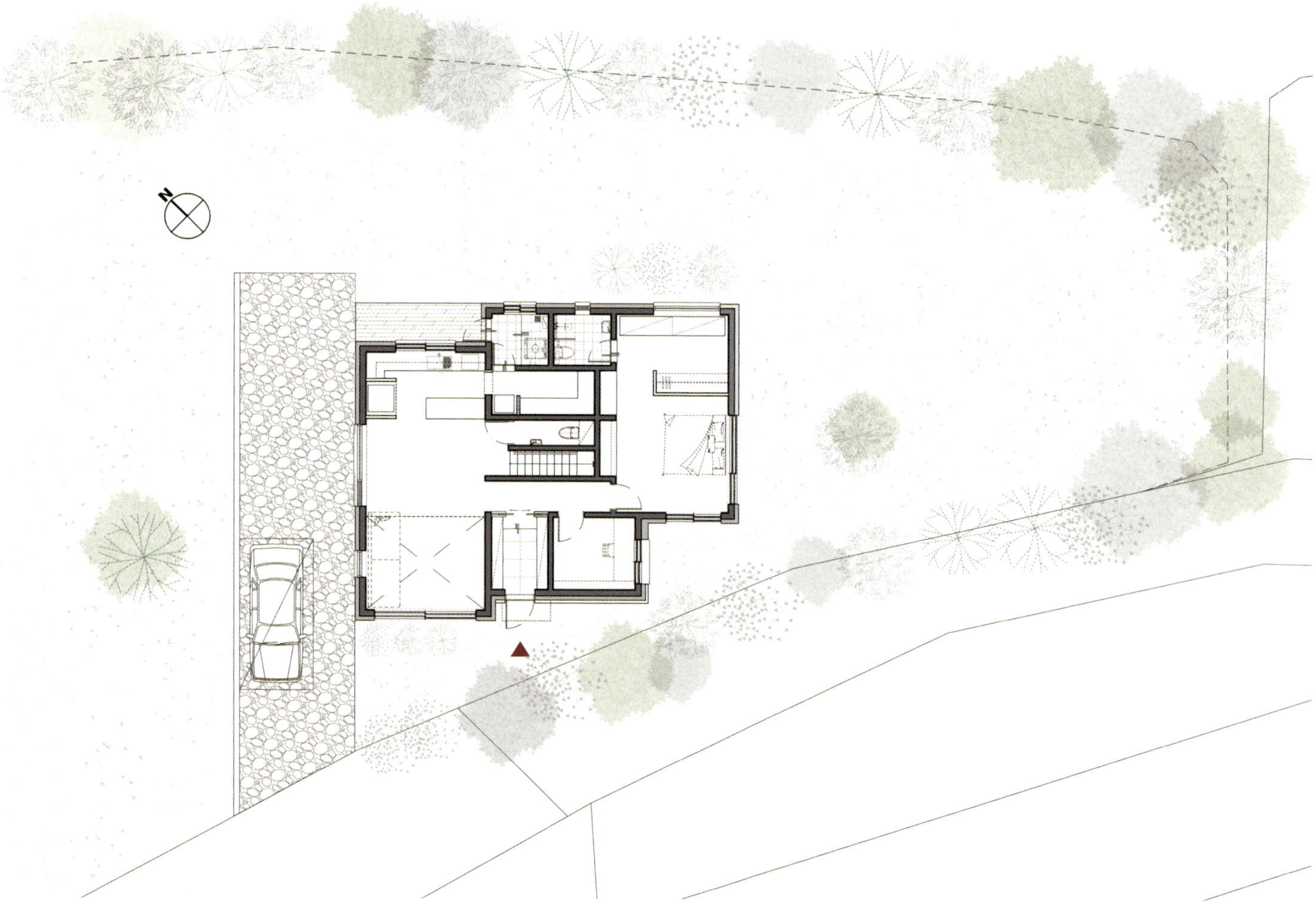

1st Floor Plan

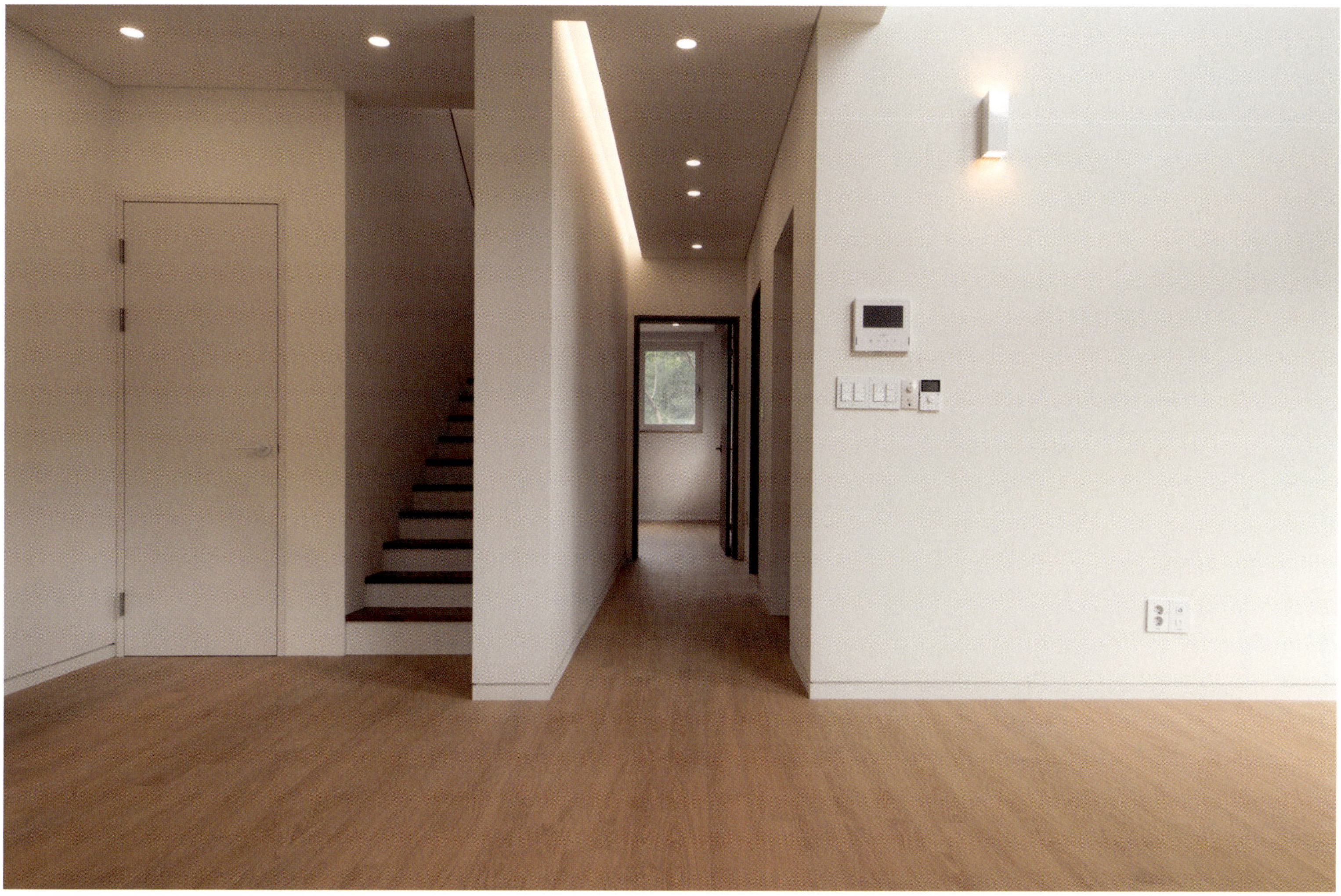

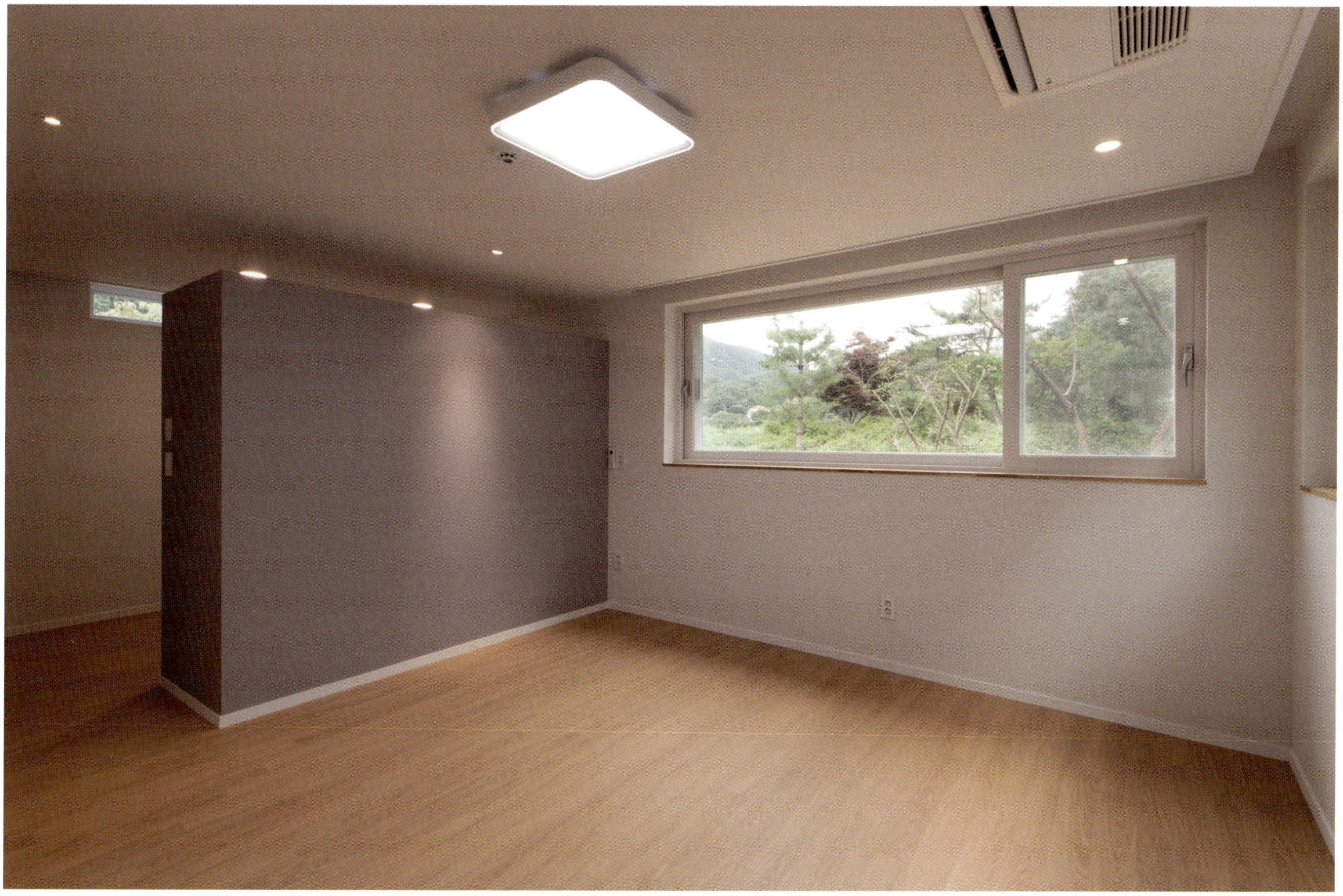

Oh U House Yangpyeong, Korea
WE Architects

Architects WE Architects / Mincheol Shin **Location** Yangpyeong, Korea **Site Area** 1,963㎡ **Bldg. Area** 116.4㎡ **Gross Floor Area** 188.42㎡ **Bldg. Scale(Floor)** 2FL **Structure** Reinforced Concrete **Exterior Finish** Roof_Zinc Panel, Wall_Red Brick, Q-Block, Deck_NewTechWood Synthetic Wood **Photos Offer** WE Architects

설계 위종합건축사사무소 / 신민철 **위치** 경기도 양평군 서종면 서후리 464-1번지 **대지면적** 1,963㎡ **건축면적** 116.4㎡ **연면적** 188.42㎡ **규모** 지상2층 **구조** 철근콘크리트 **외부마감** 지붕_징크판넬, 벽_적벽돌, 큐블럭, 데크_뉴테크우드 합성목재 **사진제공** 위종합건축사사무소

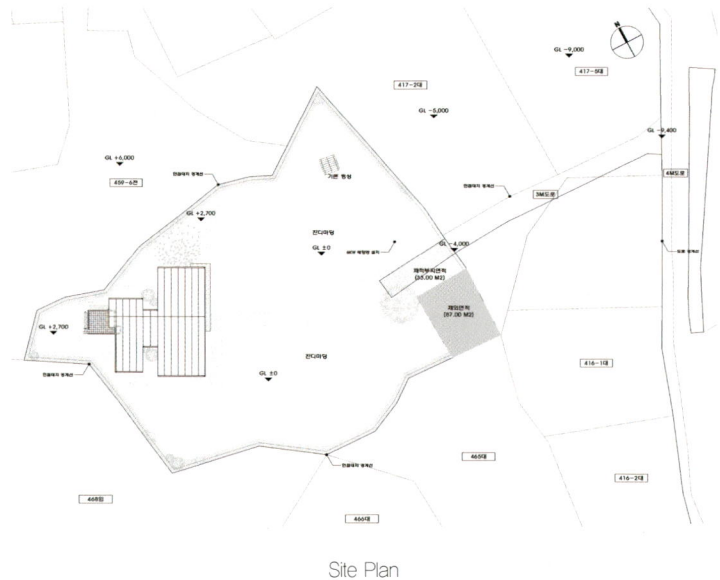

Site Plan

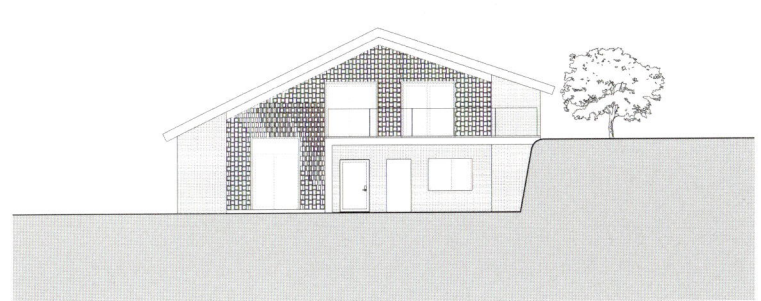
Front Elevation

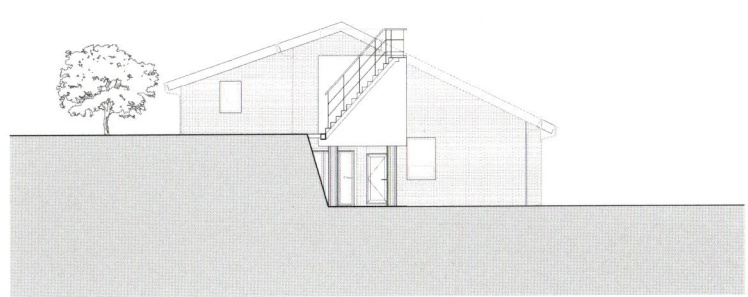
Rear Elevation

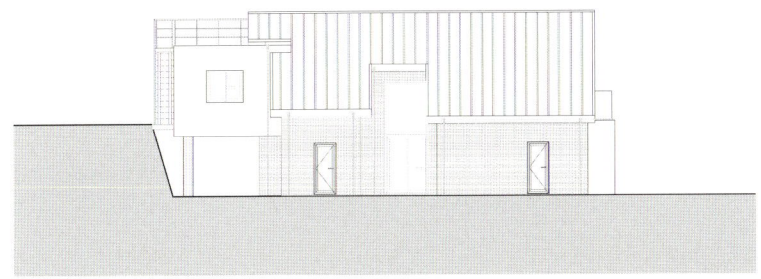
Left Elevation

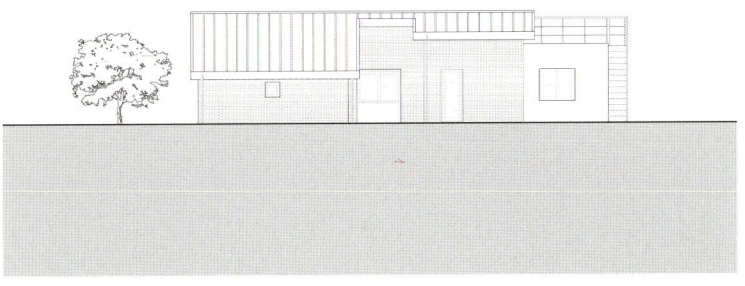
Right Elevation

The scenery seen through the rice fields under the yard of the village where they sat together over the wide yard that they saw when they first went to the site reminded me of the Jong-gatjip, which is the background of the novel land. I first envisioned using the existing house with my hands on it, and looked around the inside and outside, but the structural part, insulation, and airtightness of the windows became problems, so I decided to rebuild it. The old zelkova that seems to have been there since the village was created at the entrance of the yard, a handsome big rock in the middle of the yard, and a large ginkgo tree in the garden one step higher on the rock are so wonderful, my memory is a bit hazy, but I told the hostess, 'I didn't buy the land. You've bought time.' The design process was simple. After removing the existing house from the site, the design was carried out in the process of placing the house on nature, which had originally occupied its place. By organically combining the two-tiered manor, the first-tier part in front of the living room is expanded to become one space from the living room to the yard by minimizing the difference between the level of the site and the building. It made for a cool view. Through the balcony in front of the master bedroom on the second floor, it was easy to access the two-tiered garden on the handsome rock. The existing house, this house, and man-made structures are finite, and the original owner, nature, will remain silently as it is even after several hundred years. We borrow nature for a while, build a house, and live happily ever after. Such a case can be found in the old King grindstone magi(sales contract)rights of King Muryeong of Baekje. There are various interpretations of the magi(sales contract) right, but it describes the use of the tomb by the king of a country and the payment of land to the land god. interpreted as writing.

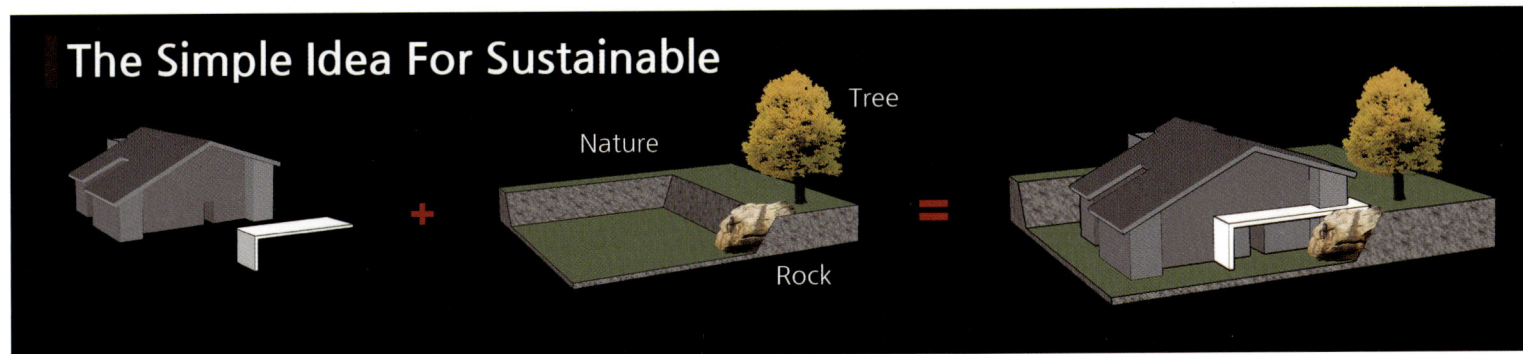

Diagram

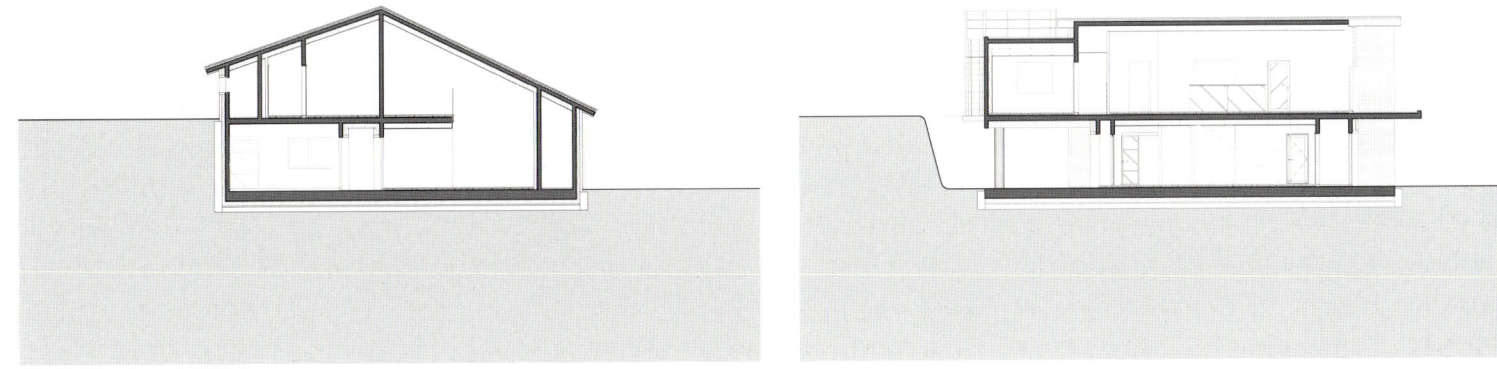

Section A　　　　　　　　　　　　　　　　Section B

오유당(吾唯堂_세월이 만든 자연과 소통 하는 씨쓰루 집)

처음 현장에 가서 보았던 넓은 마당 넘어 옹기종기 모여 앉은 마을과 마당 아래 논밭 사이로 보이는 풍경이 소설 토지의 배경이 되는 종가집을 떠오르게 하였는데 나중에 알고보니, 마을 사람들이 말하길 그 마을에서 가장 부잣집이 살았다고 하였다. 기존에 있던 집을 손봐서 사용하는 것을 처음 구상하고 안팎을 둘러보았는데, 구조적인 부분과 단열, 창호의 기밀성이 문제가 되어 다시 짓기로 하였다. 마당 입구에 마을이 생길 때부터 있었을 것 같은 오래된 느티나무와 그 마당 가운데 잘생긴 큰 바위, 그 바위 위 한단 높은 정원에 큰 은행나무가 너무 멋져서, 나도 기억이 좀 흐린데 안주인한테 '땅을 사 신게 아니라 시간의 세월을 사셨군요' 라고 하였다. 디자인 과정은 간단했다. 대지에서 기존 집을 덜어낸 뒤에 원래 자리를 차지하고 있던 자연에 집을 얹는 과정으로 디자인이 진행되었다. 2단으로 된 장원을 유기적으로 결합하여 거실 앞 1단 부분은 대지와 건물의 단 차이를 최소화하여 거실에서 마당까지 하나의 공간이 될 수 있게 확장을 하고, 거실에서 마당 넘어 느티나무가 한 눈에 들어오게 시원한 조망을 만들었다. 2층의 안방 앞 발코니를 통하여 잘 생긴 바위 위 2단의 정원으로 출입을 쉽게 하였다. 기존 있던 집이나 이번 집이나 인간이 만든 구조물은 유한하고, 원 주인인 자연은 몇 백 년이 지나도 말없이 그대로 있을 것이다. 우리는 잠시 자연을 빌려서 집을 짓고 생활을 영위하고 행복을 누리며 살뿐이다. 그러한 것은 옛 선조의 백제 무령왕 지석 매지권에서 찾아볼 수 있다. 그 매지권에 대한 다양한 해석이 있으나, 일국의 왕이 무덤을 쓰며 토지 신에게 토지대를 지불 한 것을 서술 한 것으로 현재로 보면 토지를 토지 신에게 빌려서 쓰는 것, 즉 유한한 인간이 무한한 자연에게 잠시 빌려서 쓰는 것으로 해석이 된다.

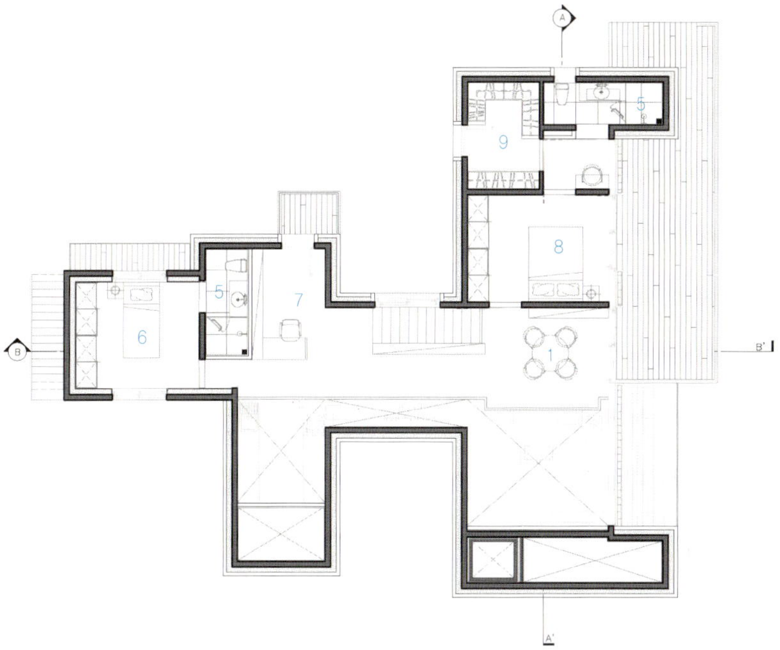

1. LIVING ROOM
2. KITCHEN & DINING
3. STORAGE
4. ENTRANCE
5. BATHROOM
6. BED ROOM
7. LIBRARY
8. MASTER BEDROOM
9. DRESS ROOM

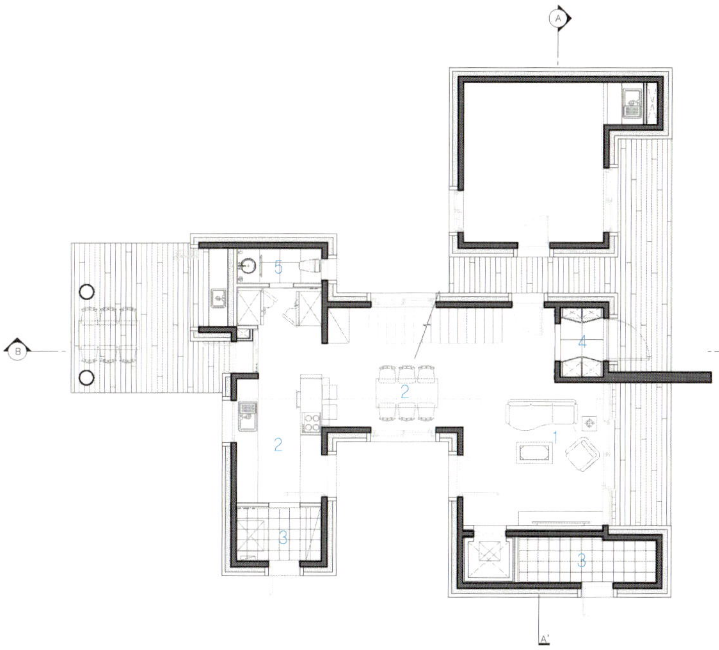

2nd Floor Plan

1st Floor Plan

Cheongdo Imdang-ri House Cheongdo, Korea

MOON ARCHITECTS

Architects MOON ARCHITECTS / Moonhyun Cho **Location** Cheongdo, Korea **Site Area** 387m² **Bldg. Area** 134.51m² **Gross Floor Area** 129.12m² **Bldg. Scale(Floor)** 1FL **Structure** Reinforced Concrete, Wooden Construction **Max. Height** 5.95m **Exterior Finish** Wall-STO, Roof-Zinc Panel, Windows-T42 Low-E Triple-Layer Glass **Photos Offer** MOON ARCHITECTS

설계 문아키건축사사무소 / 조문현 **위치** 경상북도 청도군 금천면 임당리 1855 **대지면적** 387m² **건축면적** 134.51m² **연면적** 129.12m² **규모** 지상1층 **구조** 철근콘크리트, 경량목구조 **최고높이** 5.95m **외부마감** 벽-STO, 지붕-징크패널, 창호-T42 Low-E 삼복층유리 **사진제공** 문아키건축사사무소

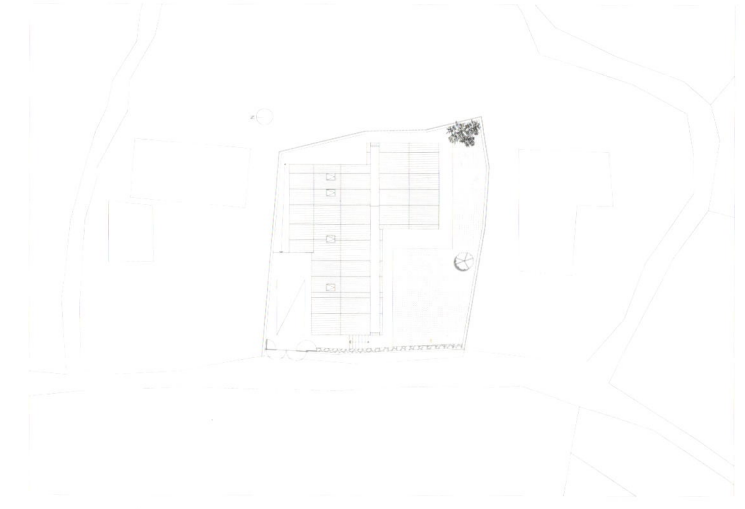

Site Plan

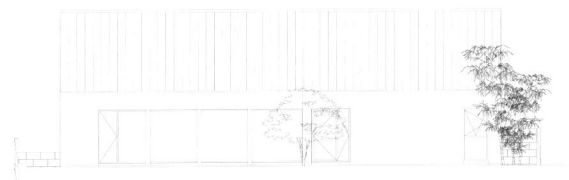

Elevation I

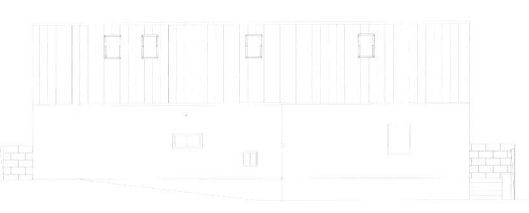

Elevation II

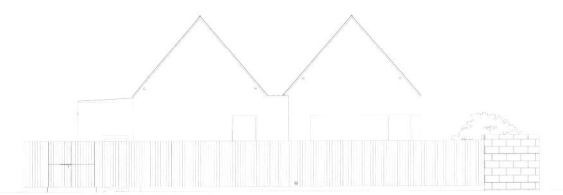

Elevation III

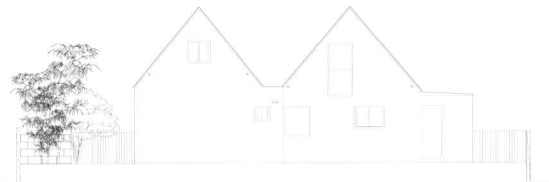

Elevation IV

He left his hometown for a living and worked tirelessly in a foreign country, so 40 years had passed. In the meantime, the two grown-up sons started their own families, and after passing the threshold of seventy, the desire to return to their hometown grew stronger. The old couple left their long-running bakery business and returned to a small town in Cheongdo, Gyeongsangbuk-do, where old memories remain intact. A thatched house that has been standing in the same place for over 100 years. After my parents died, it was almost neglected for a long time, so the abandoned house was not an environment for two people to stay. It was too old to be repaired, so we decided to demolish the old house and build a new house.

"A lot of families gather once or twice a month for family events such as ancestral rites and birthdays."

The size was determined according to the needs of the client, and a wooden house was selected for a healthy and warm house.

The building faces the same south as the existing house so that there is no inconvenience for the elderly couple, and the parking lot and courtyard are designed to be naturally separated according to the movement. In addition, as the owner's wish was 'a house where families can gather,' the interior is divided into a private area where the elderly couple resides and a public area where the family can share, and each space is assigned a role. In particular, the living room, dining room, kitchen, and attic used as public spaces have visual continuity as one space, making it a place of joyous unity even when a large family gathers.

In front of the living room is the front yard. The spacious yard is a highly utilized space that can be shared not only with family but also with neighbors. The concrete deck that surrounds the outside of the house has less maintenance and management burden than wood, so it was chosen with consideration for the convenience of the elderly couple. The canopy in front of the front door casts a shadow and provides a place to rest from the scorching sun. The public and private spaces are separated around the front entrance.

The living room, which is spacious even for the whole family, is made of solid wood such as cypress and birch plywood, which are said to be good for health considering the age of the elderly couple. Structural timber along the roof line was exposed to enrich the sense of space in the interior. The front window facing the courtyard always allows warm light to penetrate deep into the interior. The sunlight coming through the skylight acts as a soft light.

There are two sliding doors in the kitchen leading to the living room. It is a door that connects to the entrance and dining room, respectively, and is intended to reduce unnecessary movement of the elderly client and consider efficient movement. Above it is an attic made by utilizing the high floor height.

The old couple's bedroom has a window in the corner to provide light and a view of the yard at the same time. A large attic was placed above the bedroom to store various items. As it is a warehouse use, we saved space by installing a folding ladder instead of a separate staircase.

When it comes to building a house, everything is a battle against the budget. The house's goal was also to fit within the budget of a typical rural house. Currently, most of the houses newly built in the countryside are concrete or prefabricated (sandwich panel), so this house was intended to create an alternative to the farmhouse wooden house. Due to the fact that Korea is gradually changing to a subtropical climate, the roof and wall finishes were all white. White was chosen with the intention of efficiently reflecting this heat, and it worked great.

"After I built a house in my hometown, there were more family gatherings. Families who were obligated to come and go only for events now come to the pension as often as possible. Sit in the middle of the living room, watch your grandchildren, chat and share memories. Even if the guests come every week, I am happy rather than tired. Haha."

It was a project that left a lasting impression on the story of the elderly couple after moving.

청도 임당리 주택

생계를 위해 고향을 떠났고, 타지에서 허리 펼 새 없이 일하다 보니 어느덧 40년이 훌쩍 지나 버렸다. 그동안 장성한 두 아들은 각자의 가정을 꾸렸고, 일흔의 문턱을 넘기고 나니 고향 품으로 돌아가고 싶단 바람이 더욱 커졌다. 노부부는 오래 운영한 제과점 일을 내려놓고, 옛 추억이 고스란히 남아있는 경상북도 청도의 작은 마을을 다시 찾았다. 100년 넘게 그 자리, 그대로 지키고 선 초가집 한 채. 부모님이 돌아가신 이후엔 긴 시간 거의 방치되다시피 했던 터라 폐가가 된 집은 두 사람이 머물 수 있는 환경이 못되었다. 고쳐 살기에도 너무 낡아 결국 구옥을 철거하고 새로 집을 짓기로 했다.

"매달 1~2회정도 제사, 생일 등 집안 행사가 있을 때마다 가족들이 많이 모입니다."
건축주의 요구에 맞게 규모를 정하고, 건강하고 따뜻한 집을 위해 목조주택을 선택했다. 연로한 노부부가 생활하시기에 불편함이 없도록 기존 집과 같은 남향으로 건물을 배치하고, 주차장과 안마당 등이 동선에 따라 자연스럽게 구분되도록 설계했다. 또한, 건축주의 소망이 '가족들이 모일 수 있는 집'이었던 만큼 내부는 노부부가 거주하는 사적 영역과 가족이 함께 공유할 수 있는 공적 영역으로 분리하고 각 공간에 맞는 역할을 부여했다. 특히 공적 공간으로 사용되는 거실, 식당, 주방, 다락방은 하나의 공간으로 시각적인 연속성을 지녀 대가족이 모이더라도 답답함이 없는, 즐거운 단합의 장소가 되어준다.
거실 앞으로는 앞마당이 펼쳐진다. 널찍한 마당은 가족은 물론 이웃들과도 공유하는, 활용도 높은 장소이다. 집 외부를 두른 콘크리트 데크는 목재보다 유지·관리에 대한 부담이 적어 노부부의 편의를 배려해 선택하였다. 현관 앞 캐노피는 그림자를 드리우며 따가운 볕을 피해 휴식 공간을 제공한다. 정면 현관을 중심으로 공적 공간과 사적 공간이 분리된다.

가족이 모두 함께 모여도 넉넉한 거실은 노부부의 연세를 생각해 건강에 좋다는 편백나무, 자작나무 합판 등 원목을 많이 사용했다. 지붕선을 따라 구조목재를 노출시켜 실내의 공간감을 풍성하게 만들었다. 안마당을 향해 낸 전면창은 언제나 따스한 빛이 내부 깊숙이 스며들게 한다. 천창을 통해 들어온 햇살은 부드러운 조명의 역할을 해준다.
거실과 이어지는 주방에는 2개의 미닫이문이 있다. 현관, 다이닝룸과 각각 연결되는 문으로, 연로하신 건축주의 불필요한 움직임을 줄이고 효율적인 동선을 배려한 의도다. 그 위로는 높은 층고를 활용해 만든 다락이 배치되어 있다.
노부부의 침실은 코너에 적당히 창을 내어 채광과 마당 전망을 동시에 해결하였다. 침실 위로 각종 짐을 보관할 수 있는 넓은 다락을 두었다. 창고 용도이기에 별도의 계단이 아닌 접이식 사다리를 설치해 공간을 절약했다.
집을 지을 때 모든 것이 예산과의 싸움이다. 이 집의 목표 또한 일반적인 농촌 주택의 예산안에서 해결하는 것이었다. 현재 시골에 신축하는 집들은 대부분 콘크리트 또는 조립식(샌드위치 패널)이라 이 집을 통해 농가 목조주택의 대안을 만들고자 하였다. 우리나라가 아열대기후로 점점 바뀌고 있는 현상 때문에 지붕과 벽면 마감 모두 흰색으로 했는데, 실제로 집이 위치한 청도는 한여름 기온이 40℃까지 올라가는 날도 많았다. 이러한 열기를 효율적으로 반사시키려는 의도로 흰색을 선택했고, 이는 큰 효과를 보았다.
"고향에 집을 지은 후 가족들의 모임이 더 많아졌어요. 행사가 있을 때만 의무적으로 오가던 가족들이 이젠 펜션에 놀러 오듯 자주 들려요. 거실 가운데 모여 앉아 손주들을 보며 대화도 나누고 추억도 나누고. 매주 손님이 와도 힘들기보단 행복하네요. 허허."
이사 후 노부부의 이야기에 많은 여운이 남는 프로젝트였다.

1. DRAIN PLATE / THK0.9 COLOR STEEL PLATE
2. WATERPROOF SHEET ON THK18 OSB PLYWOOD
3. 2×10 RAFTER@400 / R30C INSULATION
4. THK10 GYPSUM BOARD
5. THK10 BIRCH PLYWOOD
6. 2×10 RAFTER 2PL@800 EXPOSURE
7. THK30 PINE WOOD
8. POWDER COATING ON THK12 FLAT IRON (W=50)
9. WATER-BASED PAINT FINISH ON THK21 2PL GYPSUM BOARD
10. THK0.8 SST
11. THK18 OSB PLYWOOD
12. EXTERIOR INSULATION FINISH
13. THK9 WOOD FLOORING
14. THK150 PLAIN CONCRETE / (INSTALLATION OF HOT WATER PIPING)
15. THK50 INSULATION
16. THK200 CONCRETE
17. ALUMINUM FLASHING
18. EXPOSED CONCRETE / CEMENT LIQUID WATERPROOFING
19. THK50 INSULATION
20. 2PLY OF THK0.08 PE FILM
21. THK150 RUBBLE

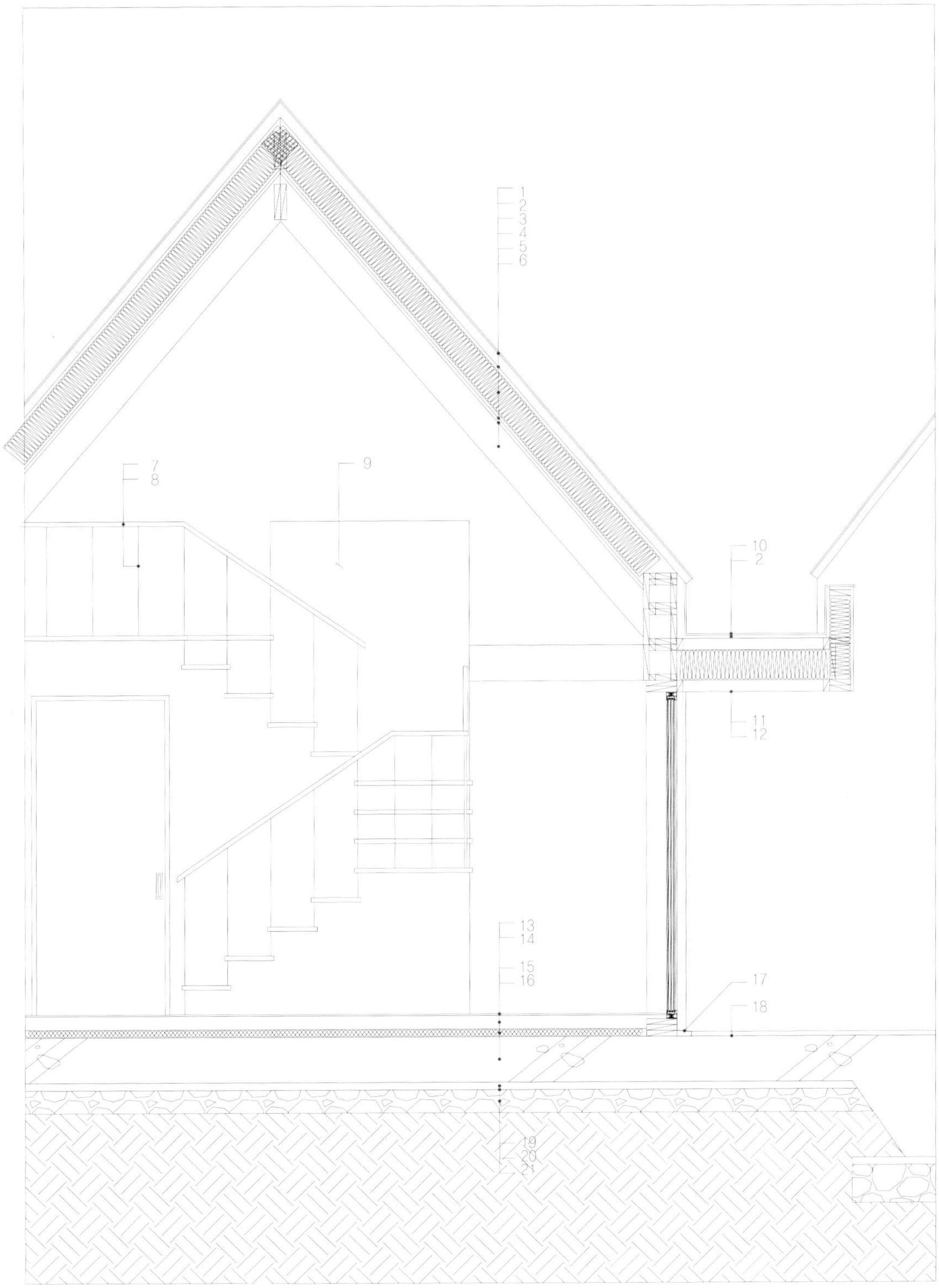

1. 배수판 / THK0.9 칼라강판 거멀접기
2. THK18 OSB합판 위 방수시트
3. 2×10 RAFTER@400 / R30C 단열재
4. THK10 석고보드
5. THK10 자작나무합판
6. 2×10 RAFTER 2PL@800 노출
7. THK30 미송원목
8. THK12 평철(W=50) 위 분체도장
9. THK21 2PL 석고보드 위 수성페인트 마감
10. THK0.8 SST
11. THK18 OSB합판
12. 외단열 마감
13. THK9 목재 플로어링
14. THK150 무근콘크리트 / (온수배관설치)
15. THK50 단열재
16. THK200 콘크리트
17. 알루미늄 후레싱
18. 노출콘크리트 / 시멘트액체방수
19. THK50 단열재
20. THK0.08 PE필름 두겹
21. THK150 잡석다짐

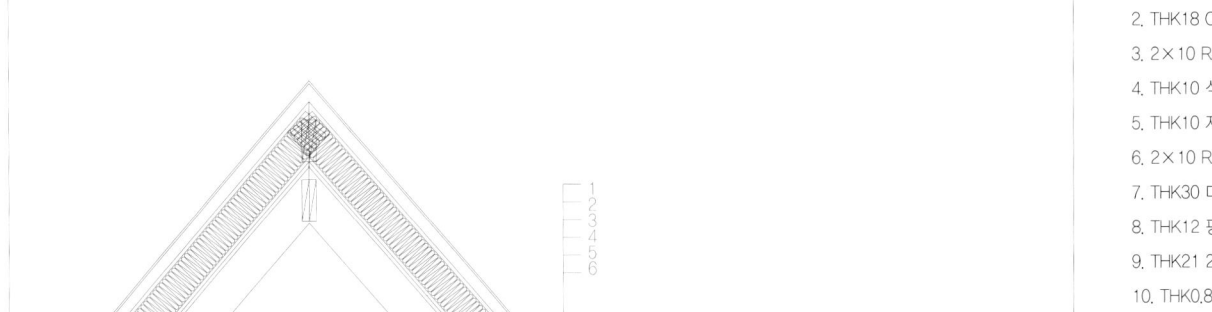

Section Detail

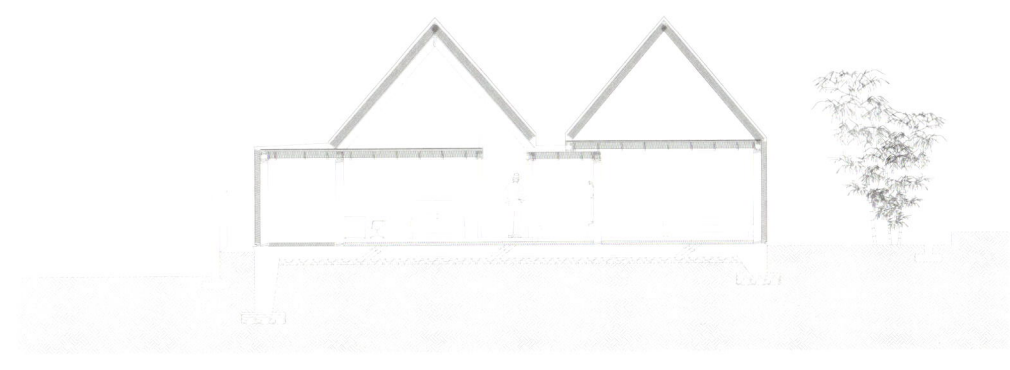

Section A

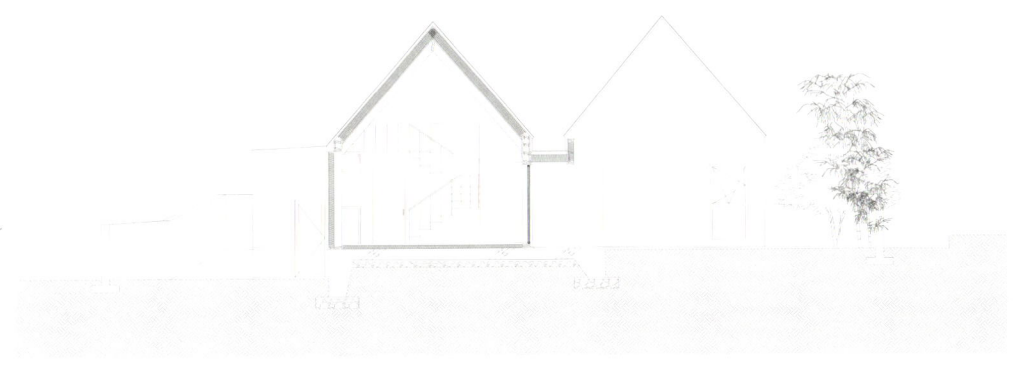

Section B

1. BEDROOM 1
2. BATHROOM 1
3. BATHROOM 2
4. ENTRANCE
5. BEDROOM 2
6. KITCHEN
7. LIVING ROOM
8. STORAGE
9. DECK
10. STORAGE / BOILER ROOM

1st Floor Plan

Attic Plan

1. ATTIC 1
2. ATTIC 2

HwaDam Byeol Seo Gimje, Korea

ilsangarchitects

Architects ilsangarchitects / Jeongin Choi, Hun Kim **Location** Gimje, Korea **Site Area** 660m² **Bldg. Area** 129.96m² **Gross Floor Area** 129.38m² **Bldg. Scale(Floor)** 1FL **Structure** Reinforced Concrete **Construction** Hesed **Client** Hyeongyook Kim **Photograhper** Seokkyu Hong

설계 일상건축사사무소 / 최정인, 김헌 **위치** 전북 김제시 공덕면 동계1길 139 **대지면적** 660m² **건축면적** 129.96m² **연면적** 129.38m² **규모** 지상1층 **구조** 철근콘크리트 **시공** ㈜헤세드 **발주처** 김형욱 **사진** 홍석규

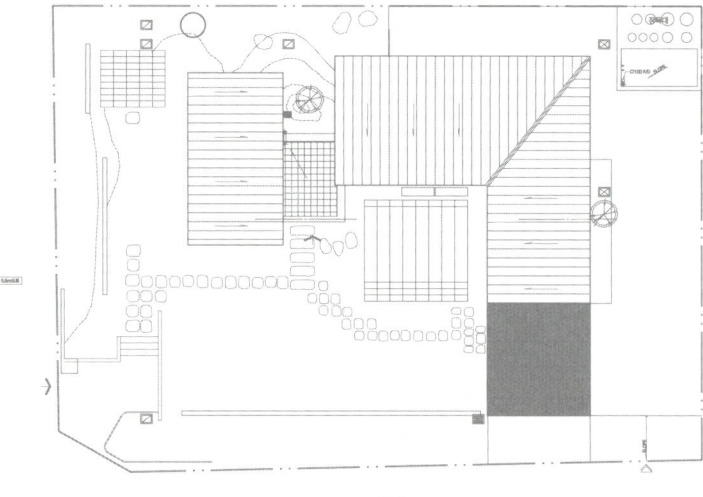

Site Plan

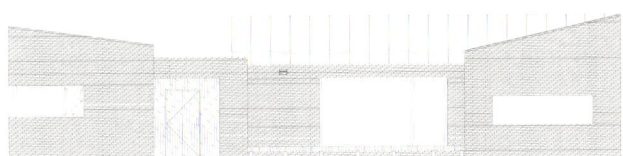

Front Elevation

Rear Elevation

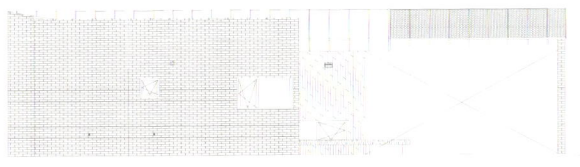

Left Elevation

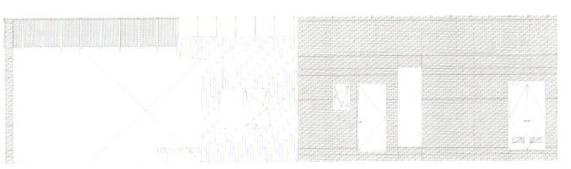

Right Elevation

The name was given by the eldest son to reflect his father's wishes. My father said it would be nice to have a space full of nostalgia, where he and his neighbors can sit on the floor, trim the waves, share sweet coffee, and talk about the lives of Cheolsu and Younghee. The basic functions are two couples and a mother. And they needed basic spaces to stay when their children visited, and a large space where many people could sit down to share food and talk flowers during holidays and family events.

My father is a lettuce farmer. He brings the lettuce harvested from the plastic house home, stores it in a low-temperature warehouse, and ships it to the public market. He wanted the flow line from the outside to the low-temperature warehouse to be simple, and he wanted the connection between the outside work space and the interior space of the residence to be close. Mother wanted the kitchen space to be adequately blocked from view rather than completely open to the living room, and she wanted a spacious utility room suitable for farm life.

Considering the characteristics of a country house, we tried to find rest, the essence of a house, as space for work and space for residence are mixed. Hwadam Byeolseo planned to properly divide the working area and residential area by building mass and closely connect them. The two areas are divided based on the guest room. It becomes the first impression of Hwadam Byeolseo, where people stay, sunlight, and wind pass through by installing a floor mat in front of the guest room. The view from the living room covers the work space with the guest room mass, allowing you to fully enjoy the landscaping of the yard and the tranquil scenery of the countryside that extends beyond the fence.

The guest room is a space for sons and a space to receive guests, so we wanted to keep a little more distance from the living room. It's like a hanok sarangchae.

At the end of the space from the outside to the inside and the view, a window was created to let the landscape and nature of the outside hang.

A large hackberry greets you in the front yard as you enter the house from the road, and when you pass through the front door and open the middle door, you are greeted by the red thorns of the rear yard, and at the end of the hallway leading to the living room, the cypress is sobbing in the sky. In the front of the living room, the scenery of the yard and the quiet farmhouse beyond the fence spreads out, and behind the living room, you can see the village through the fir trees. The guest room was designed to not only visually expand to the outside by reflecting the nature of the space, but also to go out directly to the outside through the floor. Sit on the floor and look at the four seasons of the countryside through the picture frame hanging on the wall.

Just as the toesmaru plays an important role on the outside, the indoor toesmaru is placed inside to become a symbolic space for the interior and contain various activities. Becomes a dining area, drinks her tea, reads to her granddaughter a book on her bookshelf, yawns while looking at the trees in the yard, lying on the floor and watching TV..

In order to adequately cover the complicated living in the kitchen, a wall separates the living room and the kitchen. A small opening is drilled to provide a visual connection and become a passage for food.

A corner window was planned toward the main entrance so that people entering the house from the master bedroom could be seen, and adequate shielding was given with landscaping trees.

Concept Design

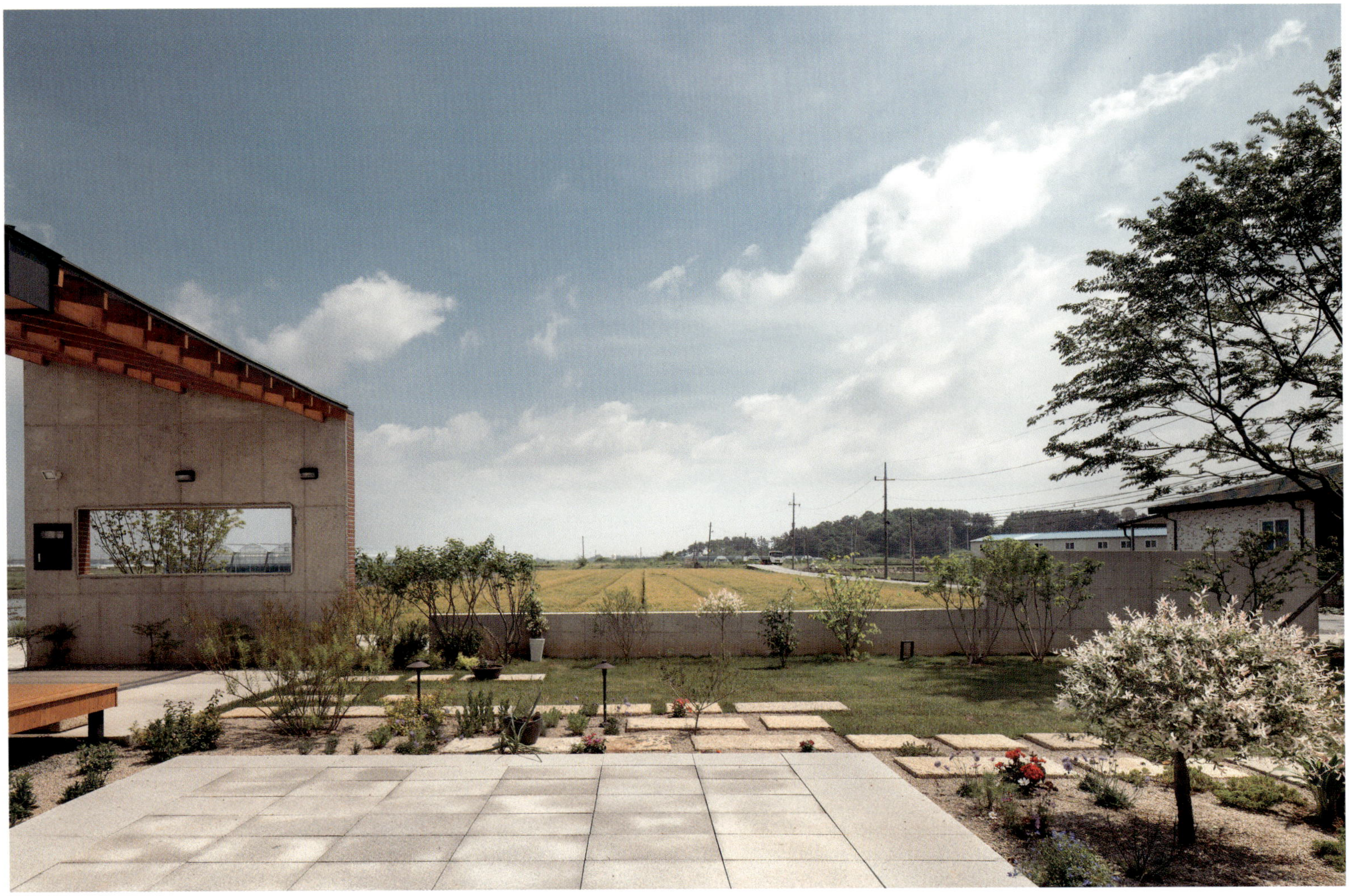

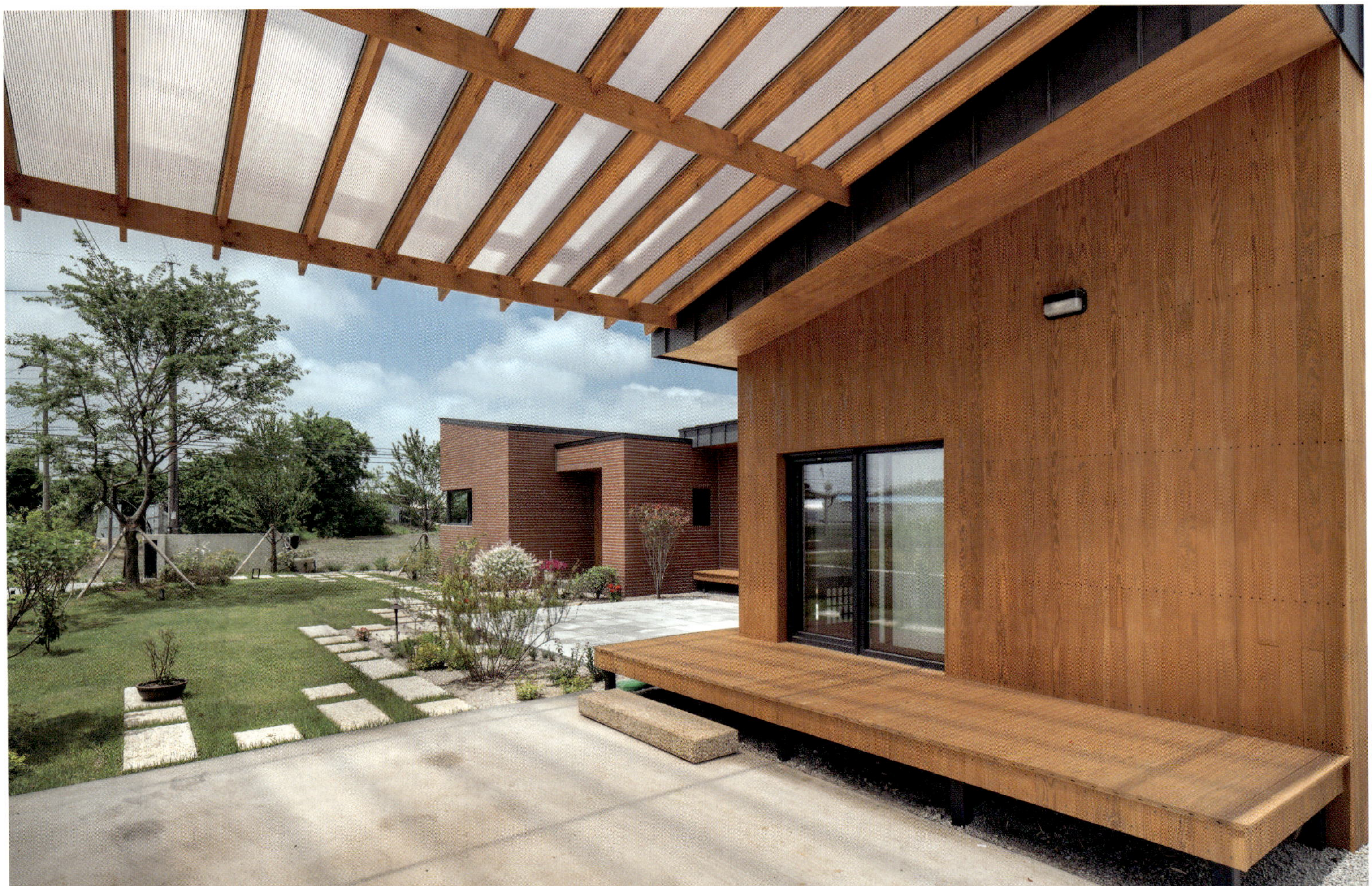

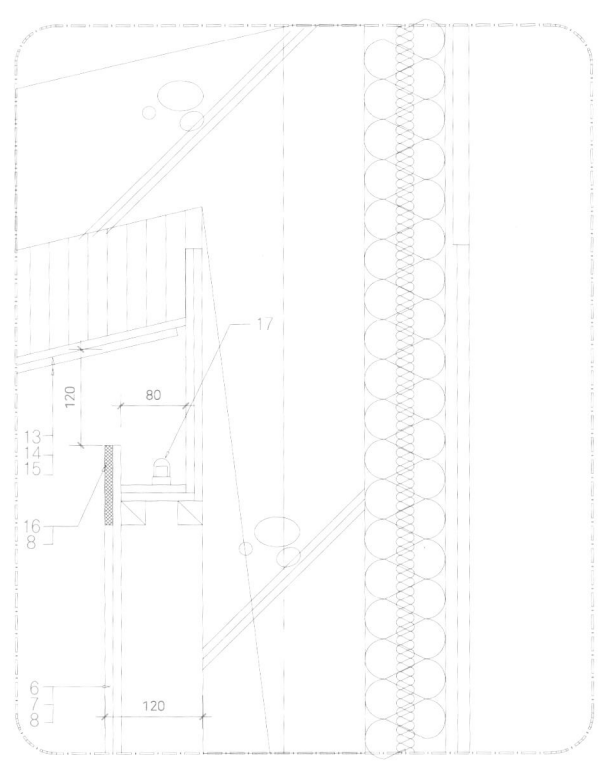

1. THK18 점토타일벽돌
2. 코킹
3. THK21 SYP탄화목재데크(무절)
   / 티쿠릴라 발티아크바 2회
4. THK30 라왕합판
   / 수성스테인후 수성바니쉬 2회
5. 목재미닫이창
6. 목재틀
7. THK20 석고보드 합지보드
8. 줄퍼티 위 친환경종이벽지
9. THK10 PVC 걸레받이
10. THK1.6 갈바름강판접기 / 조합페인트
11. THK15 탄화목재(무절) 위 오일 스테인 2회
12. THK80 경질우레탄폼단열재 2종2호
13. 목재천장틀
14. THK9.5 석고보드 2겹
15. 줄퍼티 위 친환경종이천장지
16. MDF 보강
17. T5 LED(전구색)

1. THK18 CLAY TILE BRICK
2. CAULKING
3. THK21 SYP CARBONIZED WOOD DECK(UNCUT)
   / TICURILLA VALTTI AKAVA 2 TIMES
4. THK30 RAWANG PLYWOOD
   / WATER-BASED VARNISH AFTER
   WATER-BASED STAINING 2 TIMES
5. WOODEN SLIDING WINDOWS
6. WOODEN FRAME
7. THK20 GYPSUM BOARD LAMINATED BOARD
8. ECO-FRIENDLY PAPER WALLPAPER
   ON THE PUTTY
9. THK10 PVC PLINTH
10. THK1.6 GALVANIZED STEEL SHEET FOLDING
    / COMBINATION PAINT
11. 2PLY OF OIL STAIN ON
    THK15 CARBONIZED WOOD(UNCUT)
12. THK80 RIGID URETHANE FOAM
    INSULATION MATERIAL TYPE 2 NO. 2
13. WOOD CEILING FRAME
14. 2PLY OF THK9.5 GYPSUM BOARD
15. ECO-FRIENDLY CEILING PAPER
    ON THE PUTTY
16. MDF REINFORCEMENT
17. T5 LED(BULB COLOR)

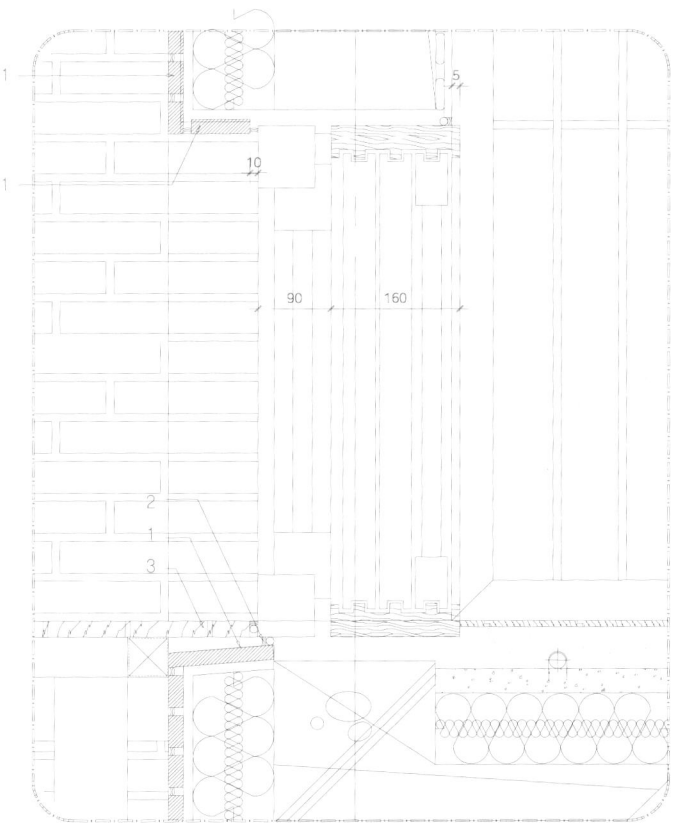

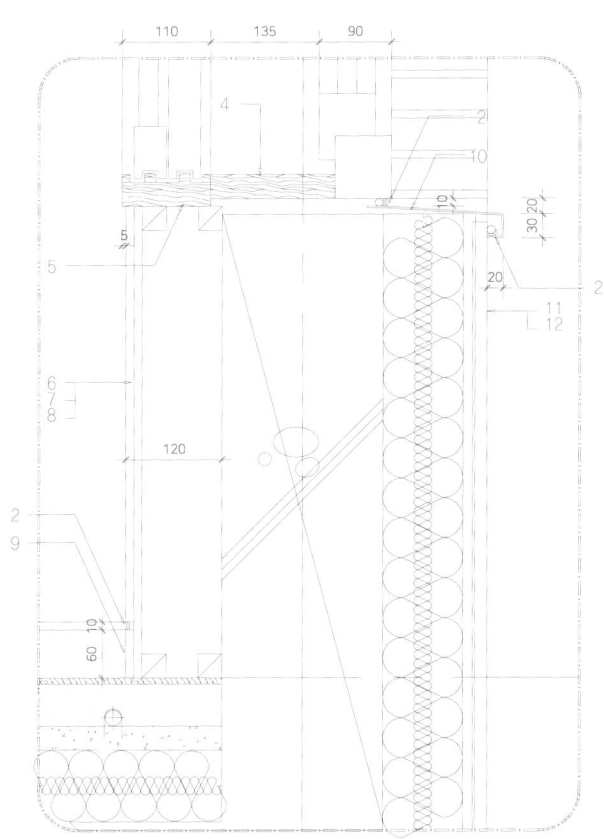

화담별서(和談別墅)

화담和談 : 정답게 주고 받는 말
별서別墅 : 농장이나 들이 있는 부근에 한적하게 따로 지은(농사를 짓는)집

아버지의 바람을 담아 큰아들이 지은 이름이다. 아버지는 동네 이웃과 툇마루에 걸터앉아 파도 다듬고, 달달한 커피도 나누고, 철수네 영희네 살림살이 이야기로 꽃을 피우며 정겨움이 넘쳐나는 공간이 있으면 좋겠다고 하였다.
기본적인 기능은 두 부부와 어머니. 그리고 자녀들이 방문했을 때 묵을 공간들이 필요했고, 그리고 명절이나 집안 행사 때 많은 인원이 함께 앉아 음식을 나누고 이야기 꽃을 피울 수 있는 대공간을 필요로 하셨다.
아버님은 상추 농사를 짓는다. 비닐하우스에서 수확한 상추를 집으로 가져와 저온창고에 보관했다가 공판장으로 출하하신다. 외부에서 저온창고로 연결되는 동선이 간결하길 원하셨고, 외부 작업공간과 주거 내부공간으로 연계가 긴밀하길 원했다. 어머님은 주방 공간이 거실과 완전히 오픈되기 보단 시야가 적절히 차단되고, 농가 생활에 적합한 넉넉한 크기의 다용도실을 필요로 했다.
전원 주택의 특성을 고려해 작업을 위한 공간과 주거를 위한 공간이 혼재 되면서 집의 본질인 쉼을 찾아주고자했다. 화담별서는 작업영역과 주거영역을 건물 매스로 적절히 나누고 긴밀히 연결되도록 계획했다. 두 개의 영역은 게스트룸을 기준으로 구분지어진다. 게스트룸 전면에 툇마루를 설치하여 사람이 머물고, 햇살이 드리우고, 바람이 지나는 화담별서의 첫인상이 된다.
거실에서 바라보는 풍경은 게스트룸 매스로 인해 작업공간을 가려주어 마당의 조경과 담장 넘어로 펼쳐진 시골의 고즈넉한 풍경을 온전히 누리게 해준다.
게스트룸은 아들들을 위한 공간이자 손님을 맞이하는 공간으로 거실에서 조금더 거리를 두고자 했으며, 이는 한옥의 사랑채와 같다.
외부에서 내부로 들어서는 공간과 그 시선의 끝에 외부의 풍경과 자연이 걸리도록 창을 내었다. 도로에서 집으로 들어서는 앞마당엔 커다란 팽나무가 반기고, 현관을 지나 중문을 열면 배면 후정의 홍가시가 맞이하고, 거실로 향하는 복도의 끝엔 오죽이 하늘하늘 흐느적거린다. 거실의 전면에는 마당과 담장 넘어의 고즈넉한 농가의 풍경이 펼쳐지고, 거실 뒤로는 전나무 사이로 마을을 바라다 본다. 게스트룸은 공간의 성격을 반영하여 외부로 시각적 확장뿐만 아니라 툇마루를 통해 외부로 직접나갈 수 있도록 계획하였다. 툇마루에 앉아 담장에 걸린 액자를 통해 농촌의 사계절을 바라다 본다.
외부에 툇마루가 중요한 역할을 하는 것처럼 내부에도 실내 툇마루를 두어 내부 상징공간이 되고 다양한 행위를 담아낸다. 식사공간이 되고, 차를 마시고, 손녀에게 책장의 책을 읽어주고, 마당의 나무를 바라보며 사색 즐기기, 마루에 누어 TV보기 등..
주방의 복잡한 살림살이를 적당히 가리기 위해 거실과 주방사이를 벽으로 구분지었다 작은 개구부 하나를 뚫어 시각적 연계를 주고 음식이 드나드는 통로가 된다.
안방에서 집으로 들어서는 이들을 볼 수 있도록 주출입구 쪽으로 코너창을 계획하고 조경수로 적절한 차폐를 주었다.
시골마을에 들어서는 집이 제 모습을 드러내기 보단 주변과 조화롭길 바랬다. 심플한 매스, 매스의 분절. 간결한 사선 지붕으로 이루어진 주택이다. '화담별서는 건축주가 지은 이름이고 일상건축이 지은 이름은 '모음집'이었다. 건물의 배치도 마당을 향해 모여지고, 집을 짓기위해 부모님과 두명의 아들들의 마음과 정성이 모여서 만들어진 집이어서 모음집이라 이름지었다.
한옥의 마당에 서서 바라다보이는 처마가 모여지는 부분이 주는 위요감. 안락함에서 건물의 디자인이 시작되었다. 외부에서 바라다 보이는 형태도 중요하지만 그 공간에서 살아가는 사람들이 느끼는 시선과 공간감을 우선시했다. 집에서 느낄 수 있는 편안함, 공간의 크기, 시야의 확장, 빛의 밝기...
화담별서는 농부의 일상을 찾아주는 일상건축의 이야기이다.

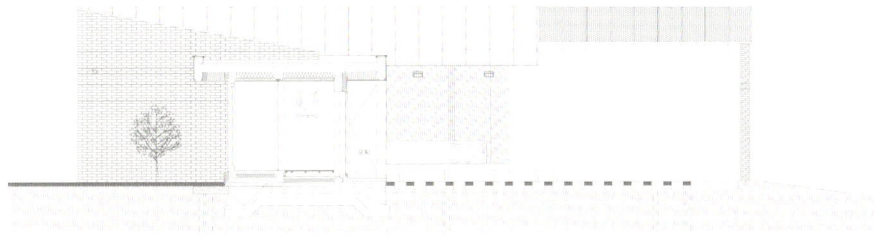

Section A

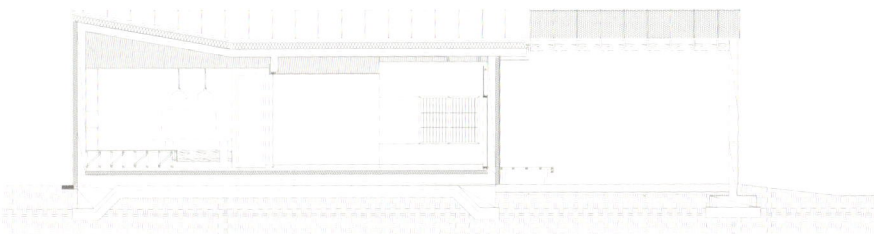

Section B

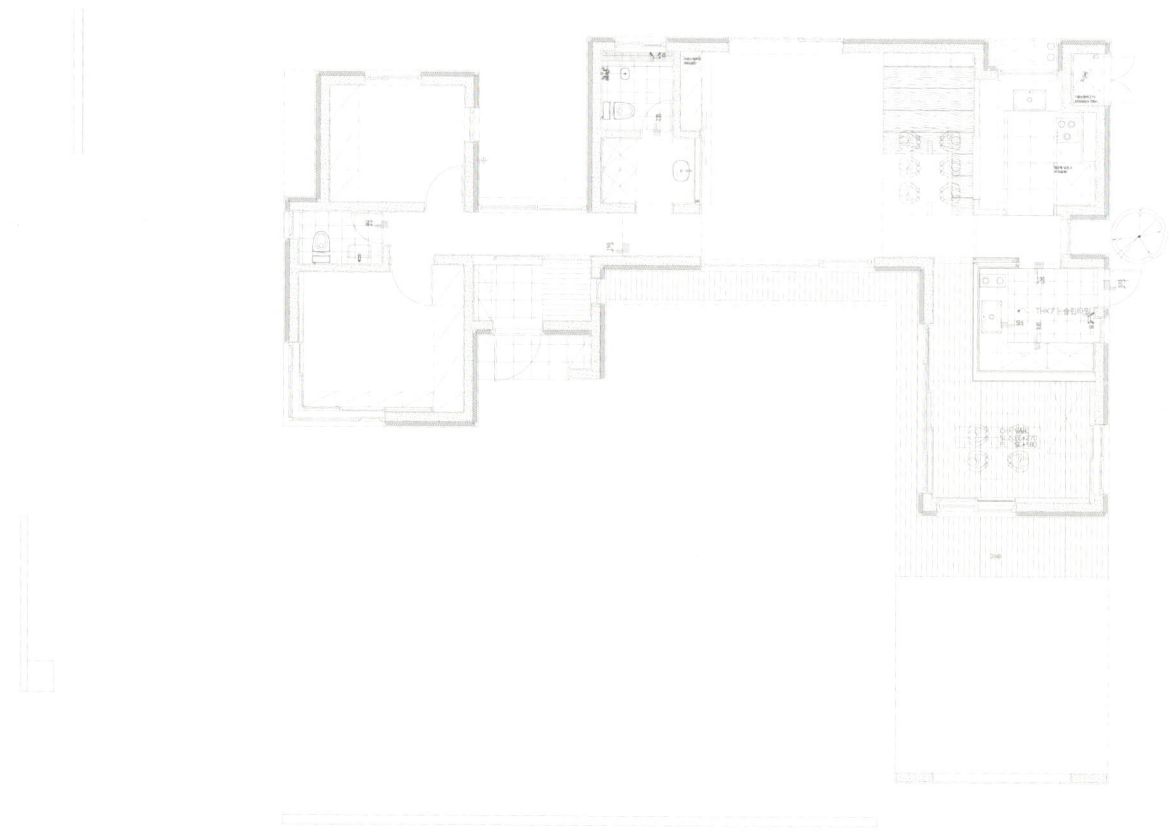

Floor Plan

127

BB4 Incheon, Korea

MaroAN Architects

Architects MaroAN Architects / Okjung Lee **Project Team** Sunyoung Lee, Seohee Ju **Location** Incheon, Korea **Site Area** 414.2㎡ **Bldg. Area** 121.67㎡ **Gross Floor Area** 297.54㎡ **Bldg. Scale(Floor)** 3FL **Structure** Reinforced Concrete **Max. Height** 12.4m **Construction** Archiload / Y. M. Yoon **Client** KRI **Photographer** Concept_ James Jeong

설계 마로안건축 / 이옥정 **설계참여자** 이선영, 주서희 **위치** 인천 서구 청라동 9-125 **대지면적** 414.2㎡ **건축면적** 121.67㎡ **연면적** 297.54㎡ **규모** 지상3층 **구조** 철근콘크리트 **최고높이** 12.4m **시공** 아키로드 / 윤영면 **발주처** ㈜케이알아이 코리아리서치 **사진 컨셉_**제임스정

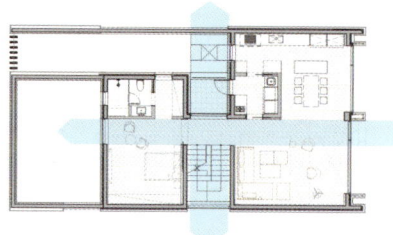

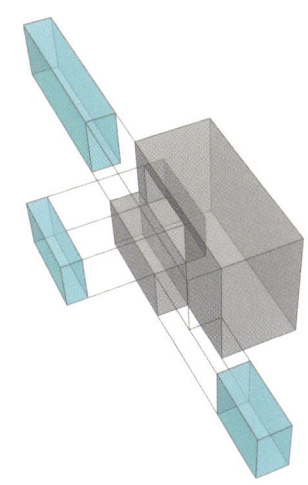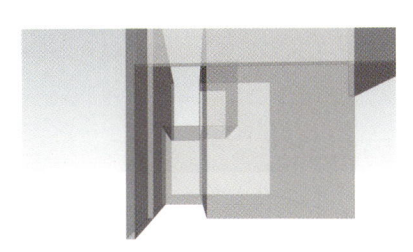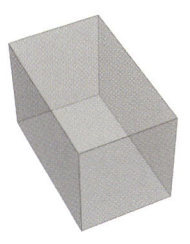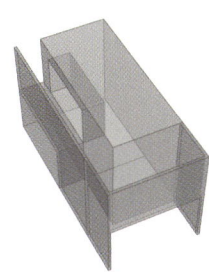

Concept Diagram

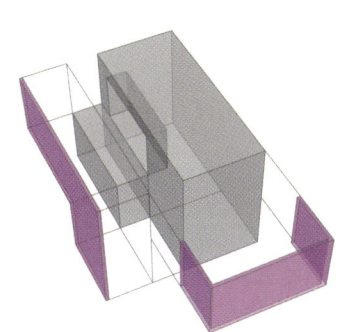

베어즈베스트 골프빌 M.NO.4

BB4 는 베어즈베스트청라 CC 단독주택단지에서 4 번째 주택프로젝트이다. 2017년 베어즈베스청라 CC 에서 첫 프로젝트를 시작으로 마로안건축의 연도별 주택시리즈물이 나오게 되었다. 각각의 프로젝트들은 대지의 형태, 면적, 주변여건 등 거의 비슷한 조건을 가지고 있고, 각 건물의 디자인은 순수하게 건축주의 생활패턴과 요구사항에 따라 방향을 달리 하고 있다. 스킵플로어집, 중정집, 필로티집 등 다양한 시도가 이루어졌다.

BB4의 클라이언트는 실험적인 건축보다는 품위있고 고급스러운 느낌, 편리한 동선, 건축적인 볼륨 등을 주로 요구하였다. 특별하진 않지만 특별한 건축이 되어야 했다. 이에 대리석 조각품을 모티브로 하여, 간결하고 군더더기 없는 디자인을 시도하였다. 편리한 동선과 조망, 건축적인 볼륨을 구축하기 위해서 페어웨이쪽으로 매스를 최대한 집중시키고, 이로 인해 다소 거대해진 매스는 도로측에 레이어드된 가벽을 활용하여 완충시켰다. 해당 지역은 지구단위계획 지침으로 인해 담장 설치가 제한 되는 등 각종 제약이 있어 도로변에 현관 및 거실이 노출되는 등 프라이버시 확보가 안 된 주택들이 다수 있었다.

이에 BB4 는 도로변측 상부 가벽으로 심리적인 경계를 구축하고, 현관으로 이르는 긴 축을 형성하여 내부로의 진입이라는 행위가 꽤 오랜시간 이루어지는 상징적인 공간을 인위적으로 계획하여 품격과 프라이버시 두가지를 동시에 확보하였다. 효율적으로 집약된 공간은 자칫 외부와의 연계에 불리할 수 있으나, 주택내에 십자축을 계획하여 외부로 확장되는 공간을 구축하였다. 긴 전이공간을 통해 현관으로 들어서게 되면 큰 창을 가진 투시형 계단실과 마주하게 된다. 큰 창 너머로 보이는 녹색의 정원이 개방감과 청량감을 주고, 계단 하단부는 계단식 벤치형태로 계획하여 단순한 동선공간이 아니라 휴게, 전시 등 다양하게 활용하게 하였다.

계단실축과 수직을 이루는 내부 복도축은 페어웨이쪽으로 열려있어 조망감과 개방 감이 확보된다. 특히 2 층 페어웨이측 전면 커튼월은 2,3 층이 오픈되어 있어서 개 방감이 극대화된다. 내부 복도축에 면해있는 거실, 방의 문도 복도폭과 동일한 사이즈의 포켓도어로 계획하여 개방감을 높였다. 또한 복도축의 조명또한 방향감을 극대화할 수 있는 라인조명을 사용하였다. 3 층 전면부를 복층형태로 보이드시킴에 따라 2 개층 보이드된 공간은 2 층거실, 2 층 식당, 3 층 거실, 3 층 테라스가 연결된 매개공간이 된다. 수직으로 일부 확장된 공간으로 인해 내부 볼륨과 공간감이 풍성해 진다. 3 층 거실은 폴딩도어로 구획된 넓은 테라스와 마주하고 있다. 폴딩도어로 인해 내외부의 경계가 흐려지고 때로는 거실의 확장공간으로, 때로는 외부테라스의 확장공간으로 활용된다.

테라스에는 드평상데크에 자쿠지 욕조를 매립하여 외부공간을 4 계절 사용할 수 있도록 하였고, 옥상공간으로 연결되는 스틸 라운드 계단은 3 층, 더 나아가 이 주택전체에서 오브제 같은 요소로 활용된다.

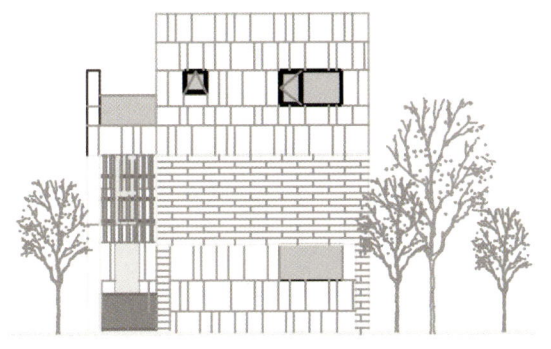

Front Elevation

Left Elevation

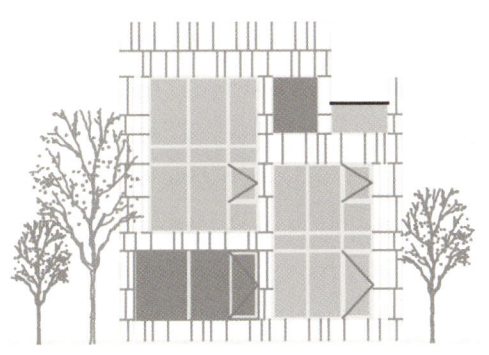

Rear Elevation

Right Elevation

BB4 is the 4th housing project in the Bears Best Cheongna CC detached house complex. Starting with the first project at Bear's Bes Cheongna CC in 2017, Maroan Architecture's yearly house series came out. Each project has almost similar conditions, such as the shape, area, and surrounding conditions of the site, and the design of each building is designed in a different direction depending on the client's life pattern and requirements. Various attempts were made, such as the Skip Floor House, the Courthouse House, and the Piloti House. Rather than experimental architecture, BB4's clients mainly requested a refined and luxurious feel, convenient circulation, and architectural volume. It wasn't special, but it had to be a special architecture. In this regard, a marble sculpture was used as a motif, and a concise and streamlined design was attempted. The mass was concentrated toward the fairway as much as possible to construct a convenient movement line, view, and architectural volume, and the somewhat huge mass was buffered by using a temporary wall layered on the road side.

Accordingly, BB4 secured both dignity and privacy by artificially planning a symbolic space where the act of entering the interior takes place for quite a long time by constructing a psychological boundary with the upper temporary wall on the side of the road and forming a long axis leading to the entrance. Efficiently concentrated space may be disadvantageous in connection with the outside, but a cross-axis was planned inside the house to build a space that expands to the outside. When you enter the front door through the long transition space, you are faced with a see-through stairwell with a large window. The green garden with a large window gives a sense of openness and freshness, and the lower part of the stairs is designed in the form of a stepped bench so that it can be used for various purposes such as resting and exhibitions, not just a simple movement space.

The inner corridor axis, which is perpendicular to the stairwell axis, is open toward the fairway, ensuring a sense of view and openness. In particular, the 2nd and 3rd floors of the front curtain wall on the 2nd floor fairway are open, maximizing a sense of openness. The living room and room doors facing the inner corridor are also designed with pocket doors of the same size as the corridor width to increase the sense of openness. In addition, line lighting was used to maximize the sense of direction for the lighting of the corridor axis. By voiding the front part of the 3rd floor in the form of a duplex, the voided space on the 2nd floor becomes an intermediate space where the 2nd floor living room, 2nd floor dining room, 3rd floor living room, and 3rd floor terrace are connected. Due to the vertically extended space, the interior volume and sense of space are enriched. The living room on the 3rd floor faces a large terrace partitioned by a folding door. Due to the folding door, the boundary between the inside and outside is blurred, and it is sometimes used as an extension space for the living room and sometimes as an extension space for the outside terrace.

On the terrace, the outdoor space can be used for four seasons by embedding a Jacuzzi bathtub in the depyeongsang deck, and the steel round staircase leading to the roof space is used as an object-like element on the third floor and furthermore, the entire house.

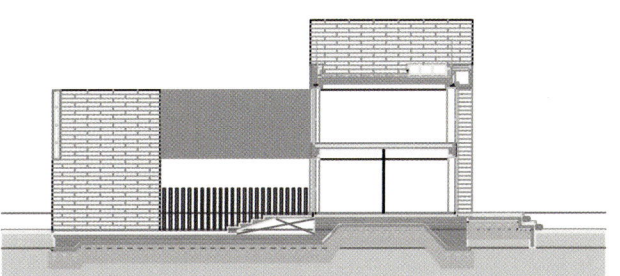

Section A

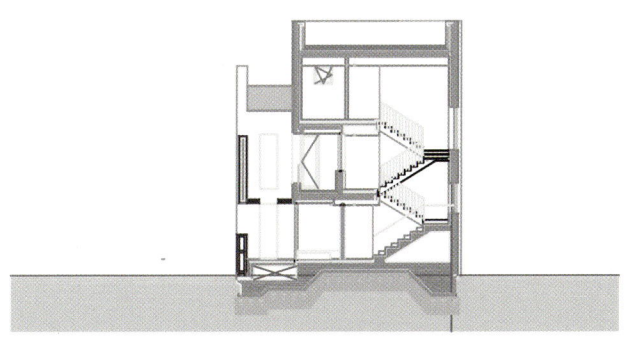

Section B

Section C

1. LIVING ROOM
2. ROOM
3. AV ROOM
4. ENTRANCE
5. BATHROOM
6. STORAGE
7. DRESS ROOM
8. TERRACE

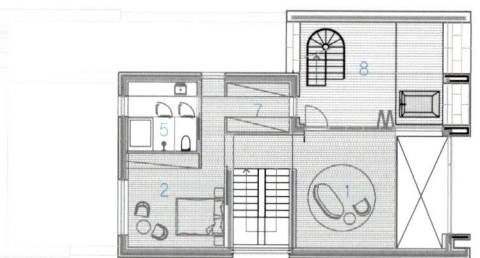

2nd Floor Plan

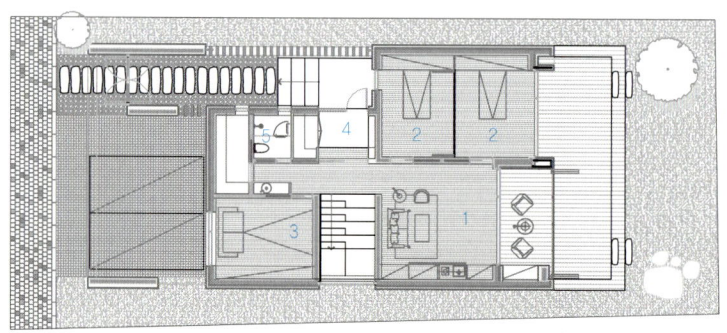

1st Floor Plan

The Sky Court House Yongin, Korea

Gyeongpiri Architecture Powerhouse

Architects Gyeongpiri Architecture Powerhouse **Project Team** Eunyoung Lee, Yuri Yoon **Location** Yongin, Korea **Site Area** 211㎡ **Bldg. Area** 120.57㎡ **Gross Floor Area** 279㎡ **Bldg. Scale(Floor)** 2FL, B1 **Structure** Reinforced Concrete **Max. Height** 8.95m **Exterior Finish** Brick, VM Zinc, Ipe Wood, U-Glass **Client** Gyeongjin Go **Photos Offer** Gyeongpiri Architecture Powerhouse

설계 경피리건축발전소 건축사사무소 **설계참여자** 이은영, 윤유리 **위치** 경기도 용인시 기흥구 **대지면적** 211㎡ **건축면적** 120.57㎡ **연면적** 279㎡ **규모** 지상2층, 지하1층 **구조** 철근콘크리트 **최대높이** 8.95m **외부마감** 벽돌, VM 징크, 이페 우드, 유-글라스 **시공** 마루디자인건설 **발주처** 고경진 **사진제공** 경피리건축발전소 건축사사무소

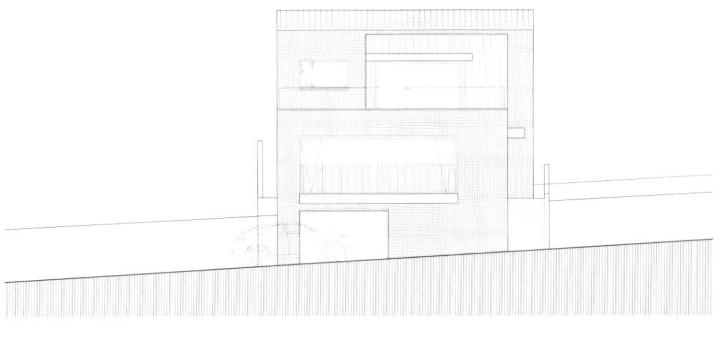

Elevation I

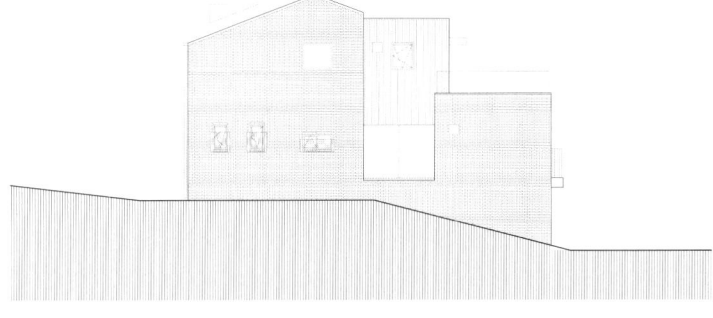

Elevation II

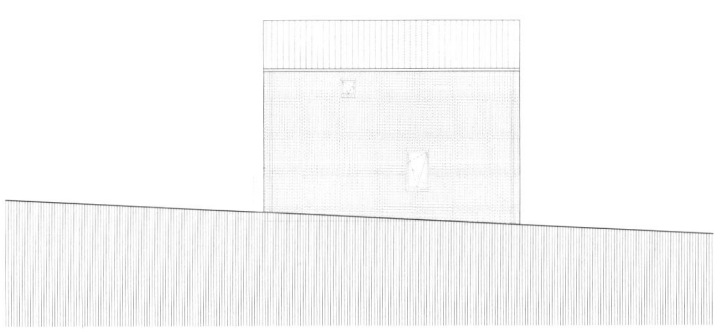

Elevation III

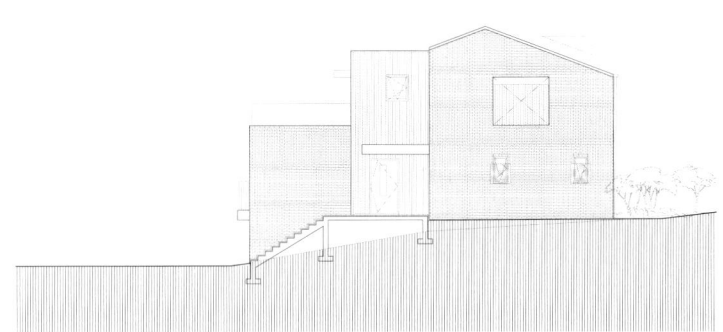

Elevation IV

Housing and design concepts

This house is completely for the daughter, built with the hearts of parents. The father wanted a hotel-like house for the daughter of a professional golfer who tours a lot abroad. The second floor of an independent space was designed as a space for his daughter. looks at the forest like the living room on the first floor.

- Design concept

The owner wanted a neat building made of bricks, and said that it would be nice to have an independent courtyard where the outside gaze was blocked. Since it is necessary to design a detached house, above all, we tried to faithfully reflect the client's needs including the requirements of the client and the experience of the architect. The core of the building's exterior is its simple personality, that is, the minimalist façade is the charm of this house.

Have you ever seen a house with a bright and friendly impression on the way? Buildings that make people happy are rare. Inevitably, we live in the mouths of buildings that look like they have been pasted on. As an architect, this is a very sad reality, and it is a problem that needs to be solved. The distance we have to see and feel is becoming obvious and loose.

So, I thought a lot about what differentiates a city building could be. As a result, it was decided to bring nature into the building. Therefore, 'the courtyard of the Sky Court House is a very important key element in a specific sense for both the owner and the owner, and is a naturalistic house that can be promoted.

A house that can give a friendly impression to passers-by and a cozy house for the family in it is the true appearance of a house that we should contemplate and create. I think that a house where inner beauty makes the family happy and outer beauty a visual entertainment for passers-by meets the ideal sense of community and the role that architecture should have in the future.

Floor plan and features of each room
- The floor plan and each room plan that reflects the client's requirements, lifestyle and environment

The sky courtyard house is quite three-dimensional. The space has both openness and protection at the same time. This is evidence that solids and voids have been properly utilized. It means that privacy and safety are guaranteed, and at the same time, it gives people a sense of security and freedom, including an open space for residents.

This house, not only in appearance but in a neat suit, is a house with an interesting space, different from what it looks like from the street. When you go up from the first floor to the second floor, the second floor hides quietly and prepares to set off firecrackers. If the first floor flows calmly with a warm and comfortable atmosphere, a different space on the second floor is preparing a surprise party. This is the jackpot of Sky Court House. As soon as you reach the second floor, the brightly shining white space makes you feel refreshed. This is a space that only the owner of the building can testify and experience, which cannot be captured in a photograph.

- Outdoor and indoor circulation plan

The kitchen, facing the living room and the courtyard space on the first floor, is both an exterior and an interior space. We made it possible for people who concentrate on housework and those who take a break in the living room to exchange glances right away, thanks to the courthouse. The courtyard is the medium and link of the family. Although the living room and kitchen are separate, they are actually partners. Communication with each other is transmitted through the courtyard, and people who live in a gaze that embraces nature naturally feel a sense of happiness. On the first floor, activities for couples and rest for the whole family are mainly held.

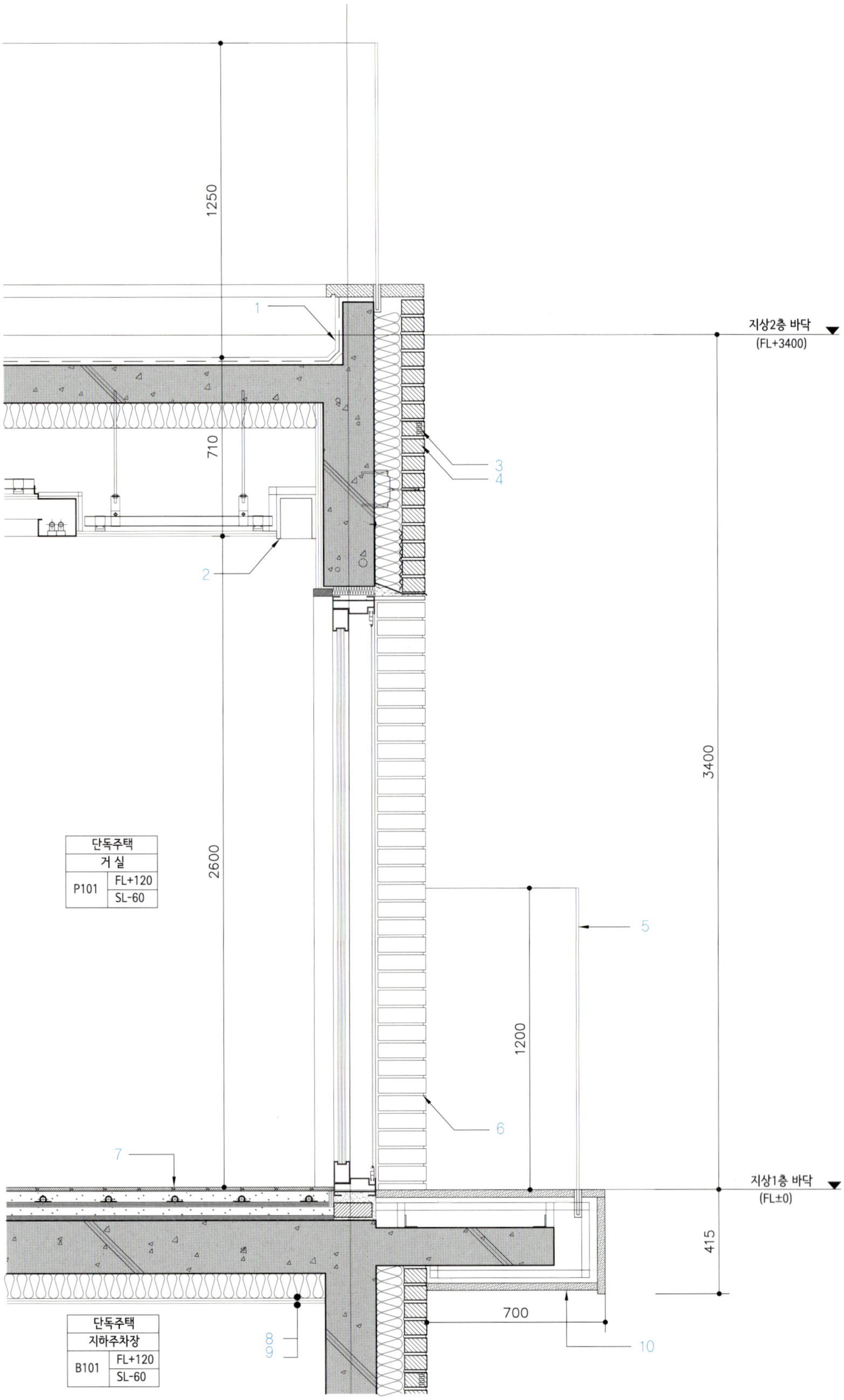

1. WATER-BASED PAINT FOR EXTERIOR USE 2 TIMES WHITE
2. THK1.6 GALVA STEEL
   COLD SAND / PUTTY / URETHANE PAINT
3. DHB HONEYCOMB DRAIN HOLE @600, HOTIZ
4. NEW EARTH STONE APP' BRICK
   0.5B STUCCO STACKING(250×90×57)
5. THK10 TRANSPARENT TEMPERED GLASS
6. THK1.6 GALVA STEEL
7. THK12 SOLID WOOD FLOORING
8. THK9.5 GYPSUM BOARD 2PLY
9. PUTTY / ACRYLIC PAINT
10. THK24 APP' MARBLE

1. 외부용 수성 페인트 2회 WHITE
2. THK1.6 갈바스틸
   한냉사 / 퍼티 / 우레탄페인트
3. DHB 하니콤 통배수구 @600, HOTIZ
4. 신토석 지정벽돌 0.5B 치장쌓기(250×90×57)
5. THK10 투명 강화유리
6. THK1.6 갈바스틸
7. THK12 원목마루
8. THK9.5 석고보드 2겹
9. 퍼티 / 아크릴페인트
10. THK24 지정 대리석

Section Detail

The balcony in front of the living room is the secret weapon for the three-dimensional shape of the house. The balcony has a view that is about the height of a person, so you can't see the living room from the outside, but it's also taking advantage of the visual fun of looking down from the living room.

Courtyard The flower of a house is the courtyard. The courtyard is the flower-like space of a modern city house. It is a space that anyone wants to install, but it requires cost, the environment of the site, and the ability of the designer. The energy radiating from the green space in the house is very passionate.

- Functions, visuals, interior effects by type of finishing material

Modern simplicity is what was planned to give the person inside a comfortable rest. LED lighting, induction installation, and eco-friendly diatomaceous earth paint on the wall were designed with in-depth consideration for the preservation of both the external environment and the residential environment.

Elevation plan

- Design features and concept

The building is composed of a gable roof of a brick building, that is, rather than pursuing a general and boring appearance, it consists of an open-roofed scalpel facing the sky, and is designed so that the upper part of the small second-floor terrace can feel nature. The large window of the living room faces the forest and is a space where no one's gaze is intruded. By focusing on the connection between the inside and the outside that cannot be felt from the outside, it is intended to solve the problem of boredom from the boredom of the overall facade.

External features related to finishing materials: The harmony between the clay bricks and the large glass in the living room is beautiful. The building adds to the fun of the facade of clear pleasure. We hope that we can shine together in the form of a quietly shining house among the surrounding complex elements.

What I want to say

- Points aimed at when designing a house

A house should be designed from the perspective of the owner or people who will live in the house, and the space should be interpreted and reflected from the perspective of the occupant. In fact, there are a lot of apartment dwellers these days, and it is very rare to find a place with a good living environment. Perhaps that's why modern people seem to think that a residence is a place where they can just stay, rather than the concept of their home. Possession of one's own entails care, affection, and concern for it. So, moreover, the house must be exclusively for the owner. Therefore, it must have the charm and characteristics desired by the architect. It is my job to help, plan and implement this. It is to make the ideal that the architect wanted and dreamed of become a reality. Architecture is not so different from human beings themselves. When he has a good side and a good side, his gaze and affection go, and he wants to make an effort to protect him. Home is the starting point of happiness, a cozy resting place, and a power plant that produces memories of my life. I think my goal as an architect is to complete a house or building with their own characteristics in color.

스카이 코트 하우스

■ 주택 및 설계 콘셉트

이집은 부모의 마음으로 짓는 온전히 딸을 위한 집이다. 해외 투어가 많은 프로골퍼 딸을 위한 호텔같은 집을 아버지는 원했다 독립적인 공간의 2층을 딸을 위한 공간으로 디자인을 했으며, 2층은 1층과 같은 중정을 품고 있으며 너른 테라스를 두고 2층의 침실은 1층의 거실과 같이 숲을 바라본다.

- 설계 콘셉트

건축주는 벽돌로 이루어진 단정한 느낌의 건물을 원했고 특별히 외부 시선이 차단된 독립된 중정이 있었으면 좋겠다고 했다. 단독주택을 디자인해야 하는 것이기 때문에 무엇보다 건축주의 요구조건과 거기에 건축가의 경험을 포함해서 건축주의 니즈(needs)를 충실히 반영하고자 하였다. 건물 외관의 핵심은 단순한 개성, 즉 미니멀한 입면은 이집만의 매력이다.

길을 가다가 밝은 느낌의 상냥한 인상을 가진 주택을 본적이 있는가?

사람을 즐겁게 만드는 건물은 흔치 않다. 어쩔 수 없이 우리는 붙여넣기를 한 것 같은 건물들의 품 아귀에서 살아가고 있다. 이는 건축가로서 상당히 참담한 현실이고, 풀어나가야 할 숙제이다. 우리가 보고 느껴야 하는 거리는 뻔하고, 루즈 해지고 있다.

그래서 본인은 도시의 건물이 지닐 수 있는 차별점이 무엇일까 많이 고민했다. 그 결과 자연을 건물에 인입하겠다는 결론을 내렸다. 그렇기에 하늘중정주택의 중정은 건축주와 본인 모두에게 특정한 의미에서 아주 중요한 핵심 요소이며 내세울 수 있는 자연주의적인 주택이다.

길에서 지나는 이들에게 상냥한 인상을 선사할수 있는 집 그리고 그속에 가족들을 위한 아늑한 집이야 말로 우리가 고민하고 만들어야 할 주택의 참 모습이 아닐까 한다. 내적 아름다움은 가족을 행복하게 하고 외적 아름다움은 지나는 이들을 위한 시각적 유희가 될 수 있는 집이 이상적인 공동체 의식과, 앞으로 건축이 가져야 할 사회적 순기능의 역할에도 부합 한다고 생각한다.

■ 평면 계획 및 각 실별 특징

- 건축주 요구 사항, 라이프스타일, 환경을 반영한 평면과 각 실 계획

하늘 중정 주택은 상당히 입체적이다. 공간은 개방성과 보호성을 동시에 지닌다. 이는 적절히 솔리드와 보이드가 활용됐다는 증거이다. 사생활과 안전은 보장됨과 동시에 거주자를 위한 개방적인 공간도 포함해 사람에게 안정감과 해방감을 고루고루 준다는 뜻이다.

겉모습만 번지르르한 집이 아니라 깔끔한 정장을 입은 이집은 길에서 보이는 모습과 다르게 공간이 재미있는 집이다. 1층에서 2층으로 올라서면 2층은 조용히 숨어 있다가 폭죽을 터뜨릴 준비를 한다. 1층이 따뜻함과 편안한 분위기로 잔잔하게 흐른다면 2층은 색다른 공간이 깜짝 파티를 준비하고 있다. 이것이 바로 하늘 중정 주택의 잭팟이다. 2층에 도달하는 순간 환하게 빛나는 화이트 공간을 보면 절로 마음이 시원해지는 느낌이 든다. 이는 사진으로는 전부 담아낼 수 없는, 건축주만이 증언하고 체험할 수 있는 공간이다.

- 실외 및 실내 동선 계획

거실과 1층의 중정 공간을 사이에 두고 마주하고 있는 주방은 외부이면서 내부적인 공간이다. 가사 일에 집중하는 사람과 거실에서 휴식을 취하는 사람이 바로 눈빛 교환을 할 수 있도록 했는데, 이는 중정 덕분이다. 중정은 가족의 매개체이자 연결고리이다. 거실과 부엌은 분리되어 있지만 실상 서로의 짝꿍인 셈이다. 서로의 소통이 중정을 통해 전해지고, 자연을 품은 시선 속에서 살아가는 사람들은 자연스레 행복감을 느끼지 않을까 싶다. 1층에서는 주로 부부의 활동과 가족 전체의 휴식이 이뤄진다.

거실 앞 발코니는 주택의 입체적 형태를 위한 비장의 무기이다. 발코니는 사람 키 높이 정도의 시선을 가지고 있어서 외부에서는 거실을 볼 수 없지만 거실에서 밖을 내려다볼 수 있는 시각적 재미도 활용하고 있다.

중정 주택의 꽃은 중정이다. 중정은 현대 도시 주택의 꽃과 같은 공간이다. 누구나 설치하고 싶지만 비용과 대지의 환경이 필요하고 설계자의 능력이 필요한 공간이다. 집 안의 녹지 공간이 뿜어내는 에너지는 몹시 정열적이다.

– 마감재 종류에 따른 기능, 시각, 인테리어 효과

모던한 단순함은 내부에 있는 사람으로 하여금 편안한 안식을 줄 수 있도록 계획한 바에 따라 이루어진 것이다. LED조명, 인덕션 설치, 벽체의 친환경 규조토 페인트 등은 외부 환경과 주거 환경의 보존을 모두 심도 있게 고려하여 디자인 한 것이다.

■ 입면 계획

– 디자인적 특징과 콘셉트

건물은 벽돌 건물의 박공지붕으로 이루어진, 즉 일반적이고 지루한 외형을 추구하기보다는 하늘을 향한 오픈지붕 형태의 메스를 구성하고 작은 2층의 테라스 상부로 하여금 자연을 느낄수있게 디자인 하였고 거실의 큰창은 숲을 향하고 누구의 시선의 침범이 없는 공간으로 외부에서 느낄수 없는 내부와 외부의 연결에 중점을 두어 전반적인 입면의 심심함으로부터 벗어나 지루함의 난제를 유쾌하게 풀고자 했다.

마감재 관련한 외형적 특징: 점토 벽돌과 거실의 큰 유리 의 조화는 아름답다. 건물은 명쾌한 즐거움이라는 입면의 재미를 더한다. 주변의 복잡한 요소의 주택들 속에서 얌전히 빛나는 주택의 모습으로 함께 빛날 수 있기를 희망한다.

■ 인테리어 콘셉트

– 인테리어 콘셉트와 특징(현관, 거실, 중정, 주방, 침실)

모든 공간은 거주자 입장에서 사용성이 좋고 몸에 좋은 집으로 계획되어야 한다. 본인은 심플하고 단순한 인테리어와 친환경자재를 계획하였으며 거실의 공간은 보다 편하고 자유롭게 이용할 수 있도록 구성했다. 탁 트이고 햇살이 따사롭게 들어오는 거실의 포근함은 이루 말할 수가 없다. 그리고 주방은 중정을 향유하는 공간에 어울리게 아일랜드 주방을 배치하였고 벽은 다수의 수납을 고려하여 공간들을 설계하였다.

■ 전하고 싶은 말

주택 설계 시 지향하는 점: 주택은 건축주, 혹은 그 집에 살게 될 사람들 입장에서 설계하고 거주자 입장에서 공간을 해석하여 반영해야 한다. 사실 요즘은 아파트 거주자가 많아지고, 주택의 거주 환경이 좋은 곳은 몹시 드물다. 그래서일까 현대인들은 거주지가 자기의 보금자리라는 개념보다는 그냥 머물 수 있는 곳이라고 생각하는 것 같다. 인간에게 있어 자신의 것이라는 소유욕은 그에 대한 관리, 애정, 관심을 수반한다. 그래서 더욱이 집은 오로지 건축주를 위한 것이어야 한다. 그렇기에 필히 건축주가 원하는 매력과 특징을 지니고 있어야 한다. 이를 도와주고 계획하고 실천하는 것이 나의 일이다. 건축주가 원하고 꿈꿔왔던 이상을 현실로 펼쳐주는 것 말이다. 건축은 인간 그 자체와 별반 다르지 않다. 좋아하는 면모와 장점이 있어야 눈길과 애정이 가고 그를 지키기 위한 노력을 하고 싶어진다. 집은 행복의 시작점이고, 아늑한 휴식처이고 내 인생의 추억을 생산하는 발전소다. 그들만의 특성을 지닌 주택, 혹은 건물을 색깔 있게 완성해주는 것이 건축가로서 지니고 있는 나의 지향점이라고 생각한다.

1. ROOF TERRACE
2. FAMILY ROOM
3. ROOM 2
4. LIBRARY
5. TERRACE
6. BATH ROOM
7. DRESS ROOM

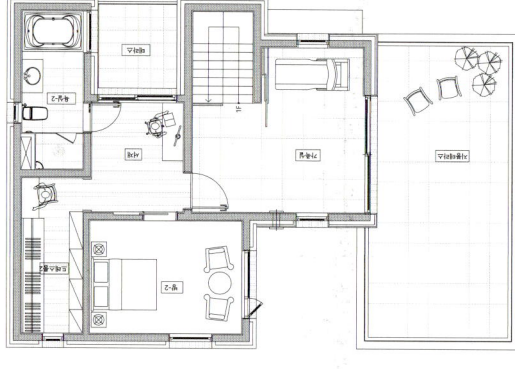

2nd Floor Plan

1. LIVING ROOM
2. KITCHEN & DINING
3. ROOM 1
4. DRESS ROOM
5. COUPLE'S BATHROOM
6. ENTRANCE
7. SHOE CLOSET
8. TOILET
9. LAUNDRY ROOM
10. SUB KITCHEN

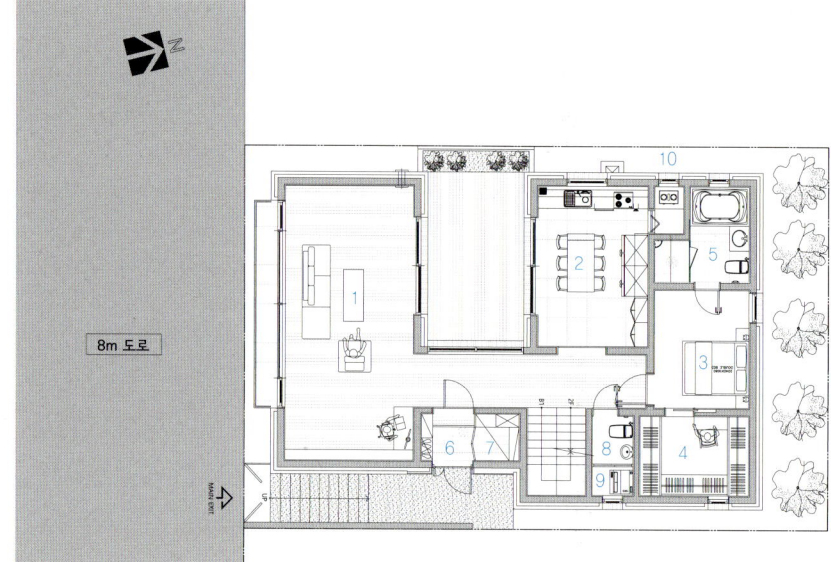

1st Floor Plan

1. EXERCISE ROOM
2. WINE STORAGE
3. PARKING LOT

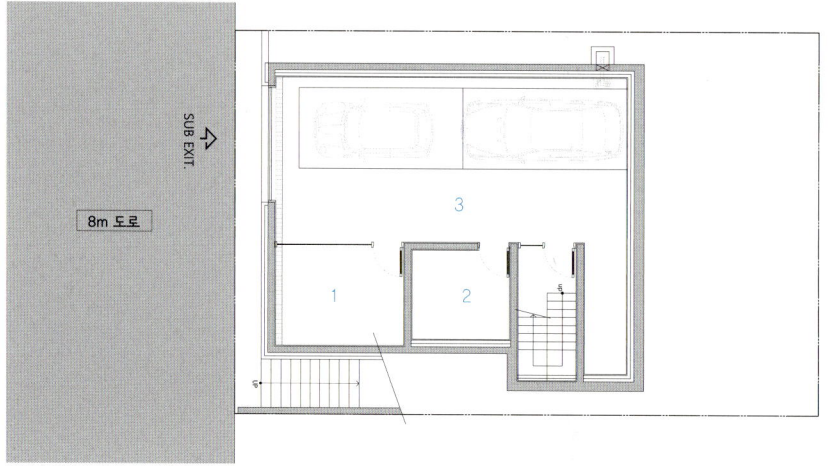

Basement Floor Plan

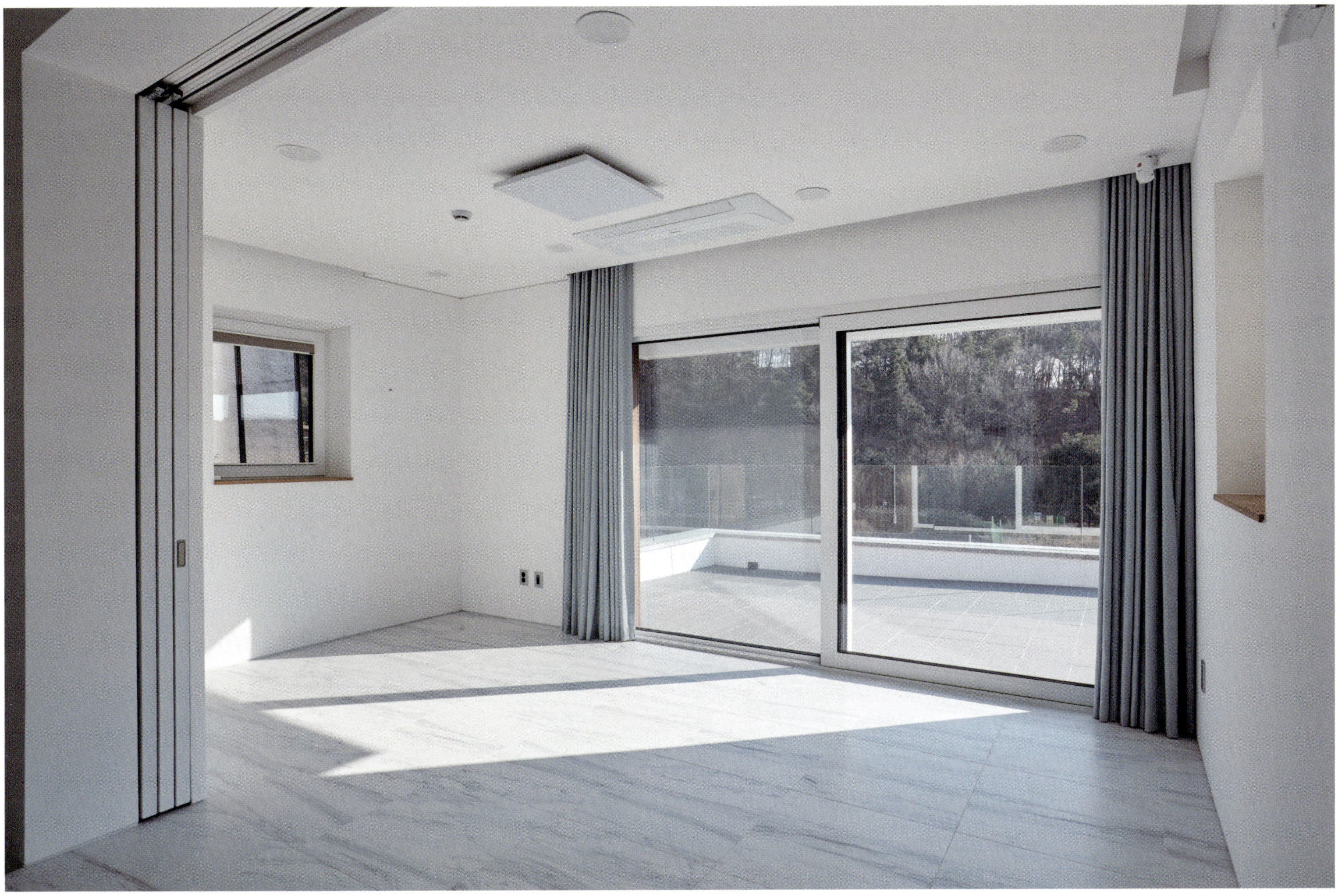

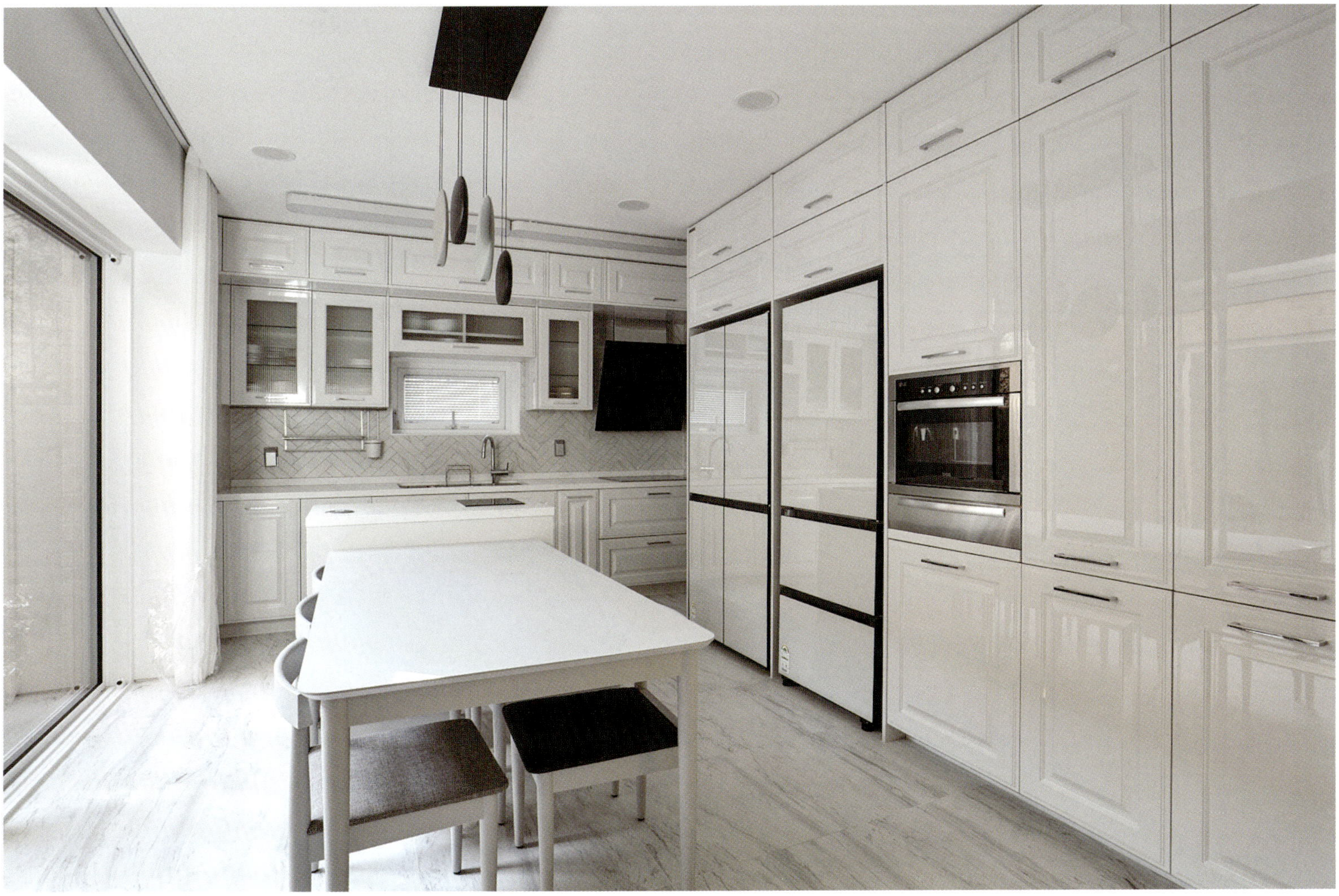

House on hill by the sea Goseong-gun, Korea

JNPeople Architects

Architects JNPeople Architects / Sehwan Jang, Seonghee In **Location** Goseong-gun, Korea **Site Area** 439㎡ **Bldg. Area** 155.46㎡ **Gross Floor Area** 35.41㎡ **Bldg. Scale(Floor)** 1FL **Structure** Reinforced Concrete **Max. Height** 6.97m **Exterior Finish** Clay Brick **Photos Offer** JNPeople Architects

설계 제이앤피플 건축사사무소 / 장세환, 인성희 **위치** 강원도 고성군 백촌리 94 **대지면적** 439㎡ **건축면적** 155.46㎡ **연면적** 35.41㎡ **규모** 지상1층 **구조** 철근콘크리트 **최고높이** 6.97m **외부마감** 점토벽돌 **사진제공** 제이앤피플 건축사사무소

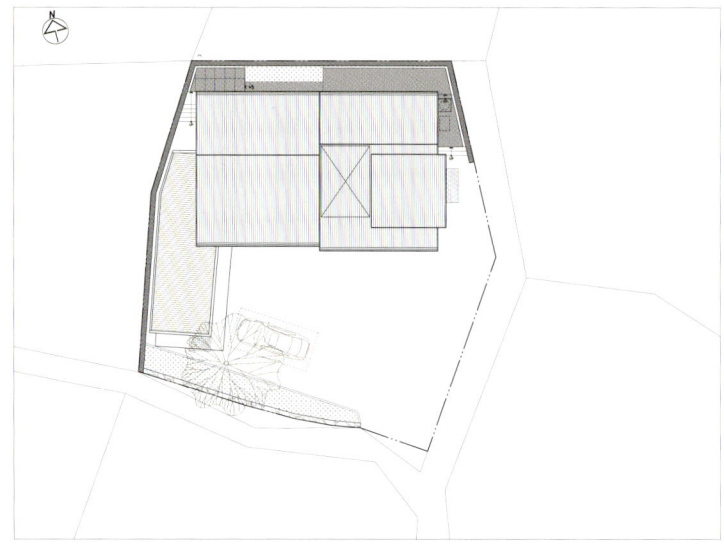

Site Plan

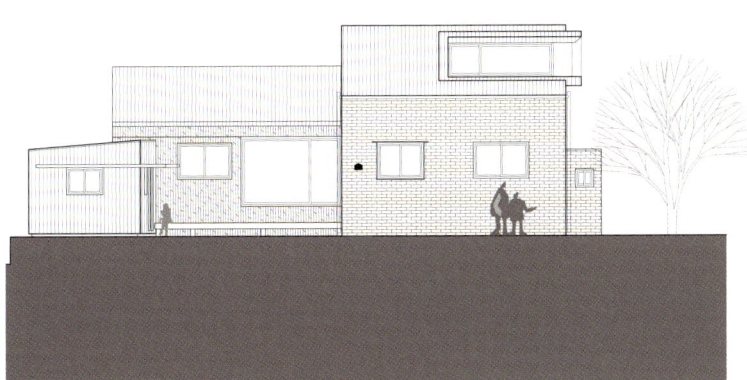

Front Elevation

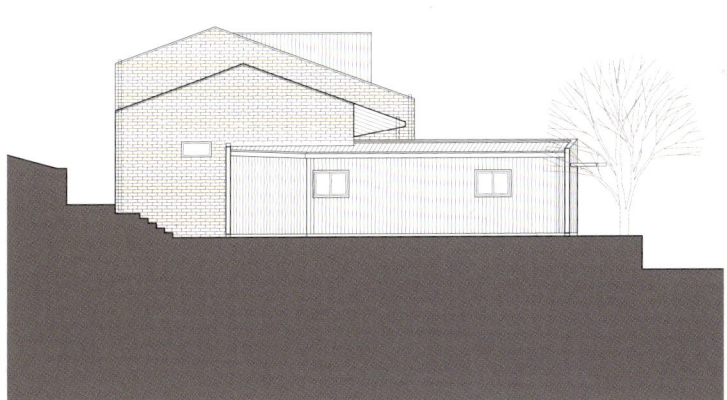

Left Elevation

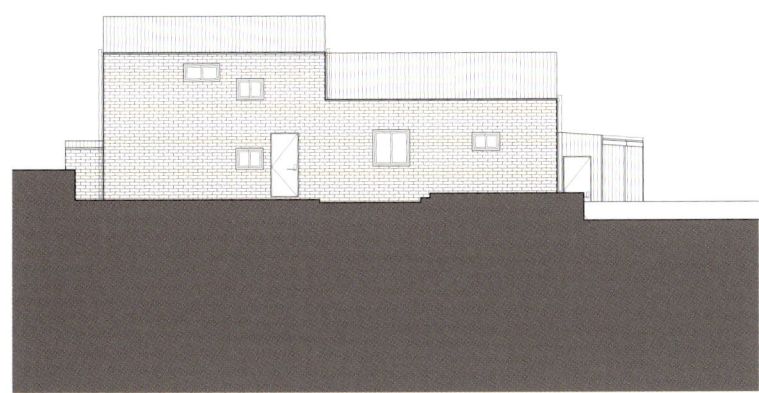

Rear Elevation

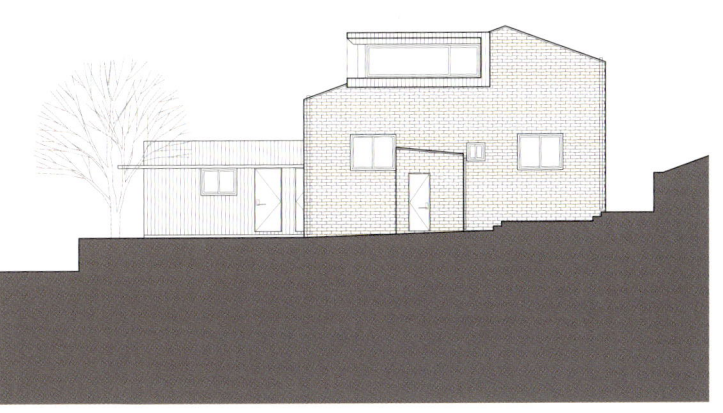

Right Elevation

It was in the house where my parents had been living for a long time on a hill overlooking the sea off Ayajin, Goseong-gun, Gangwon-do. The parents were not very positive about the new construction, but their son felt uncomfortable in life and decided to do the new construction.

Farmhouse houses have several meanings. Although it has the meaning of a house in a rural area or a house that receives support, the definition of "a house that contains rural life and the long life of elderly parents" seemed to be the most architectural. In rural areas where there is no choice but to do a lot of agricultural activities outside, agricultural tools and grain sheds, outdoor toilets, outdoor faucets, and long docks are essential functions(programs). In that respect, I thought it was important to clarify the programs(functions) needed for the house and to organize the space containing them clearly. But sometimes, familiarity with my body is more important than function or convenience. Spatial composition, shapes and materials that we inadvertently refer to are also used and made for functional needs, but they seem to blend with our lives and become memories with time. So, as time passes and it deteriorates, it has to be rebuilt again for comfort, but we tried to maintain the emotional part of the memories as much as possible.

The long eaves seen in hanok and single-story houses and the tall floor that you meet after climbing one tier can be said to be functional and emotional appearances and spaces that show traces of long life. In addition, the old movement line used by parents such as the persimmon tree contained in the yard, the height of the living room on the first floor overlooking it, the entrance shape, and the location of the warehouse were maintained as much as possible if there were no major inconveniences.

In addition, on the 2nd floor level of the hill, the sea off Ayajin was visible, and using that point, a spacious attic with sea views was prepared.

1. COLOR STEEL SHEET SEAM
2. T150 RIGID URETHANE FOAM
3. ADJACENT SITE BOUNDARY LINE
4. UNEXPOSED GUTTER
5. T50 RIGID URETHANE FOAM
6. WALL BRICK DECORATION STACKING(JOINT COLOR : GRAY)
7. DRAINAGE
8. DRIP
9. RED CEDAR STICKING VERTICALLY
10. T100 RIGID URETHANE FOAM
11. T60 LEVELING CONCRETE
12. PE FILM 2PLY
13. T200 RUBBLE

1. 칼라강판 거멀접기
2. T150 경질우레탄폼
3. 인접대지 경계선
4. 비노출 거터
5. T50 경질우레탄폼
6. 벽고벽돌 치장쌓기(줄눈컬러 : 회색)
7. 배수로
8. 물끊기
9. 적삼목 세로붙임
10. T100 경질우레탄폼
11. T60 버림콘크리트
12. PE필름 2겹
13. T200 잡석다짐

Exterior Wall Detail

바닷가 언덕위 집

강원도 고성군 아야진 앞바다가 보이는 언덕 위에 부모님이 오랫동안 생활해 오던 주택이 있었다. 부모님께서는 신축에 대해 매우 긍정적이지는 않았으나 생활에 불편함을 느낀 아들이 신축을 결정하였다.

농가주택은 여러 의미를 담고 있다. 농촌에 있는 주택 혹은 지원을 받는 주택 등의 의미가 있지만 '농촌의 삶 그리고 노부모님의 오랜 삶을 담은 주택'이라는 정의가 가장 건축적인 것 같았다. 바깥에서 많은 농업활동을 할 수밖에 없는 농촌에서는 농기구 및 곡물 창고, 바깥 화장실, 옥외수전, 장독 등은 필수 기능(프로그램)이다. 그런 면에서 집도 필요한 프로그램(기능)을 명확히 하고 그것을 담는 공간을 담백하게 구성하는 것이 중요하다고 생각했다. 하지만 때로는 기능이나 편리함보다 내 몸에 익숙한 것이 더 중요할 때도 있다. 공간 구성. 우리가 무심코 참조하는 조형, 재료들도 기능적 요구에 의해 사용되고 만들어지지만 어느덧 우리의 삶과 어우러져 시간과 함께 추억이 되는 것 같다. 그래서 시간이 지나고 노후화 되면서 쾌적함을 위해 다시금 새로이 지어질 수 밖에 없지만. 추억이 담긴 감성적인 부분은 최대한 유지 하기 위해 노력했다.

한옥 및 단층주택에서 보여지는 긴 처마와 한 단 올라선 뒤 만나는 툇마루 등은 오랜 생활의 흔적이 묻어나는 기능적이면서도 감성적인 모습과 공간이라 할 수 있다. 그밖에 마당에 담겨진 감나무. 그것을 내려다 보는 1층 거실 높이, 현관진입 형태. 창고위치 등 부모님이 사용하던 오랜 동선과 그 동선에서 보여지는 여러 모습들도 큰 불편이 없다면 최대한 유지하였다. 아울러. 언덕 2층 레벨에서는 아야진 앞바다가 보이는 상황이었고 그런점을 활용해 바다조망이 가능한 널직한 다락을 마련하였다.

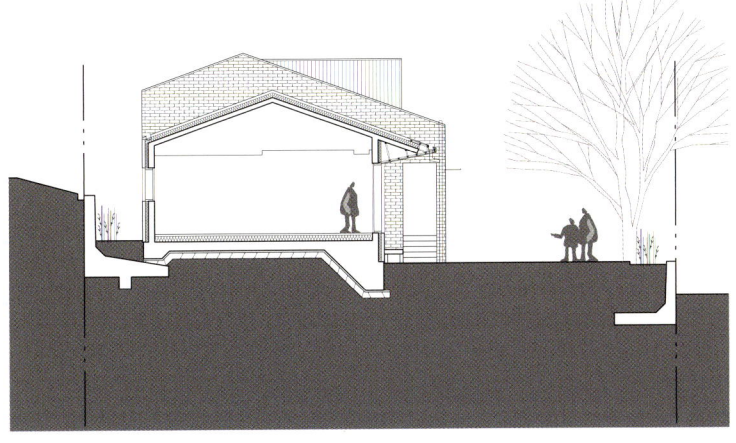

Longitudinal Section

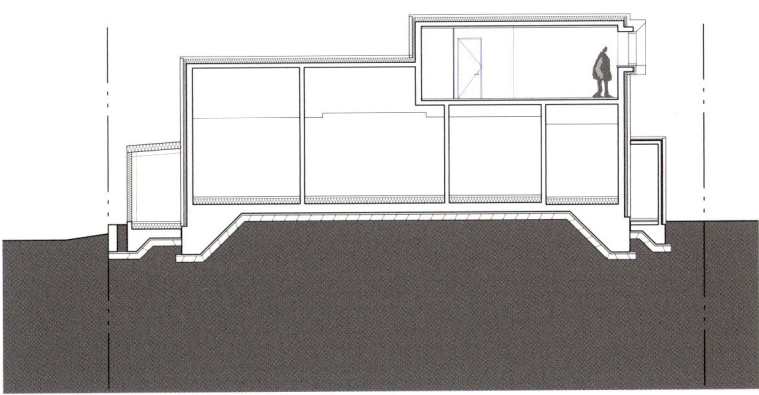

Cross Section

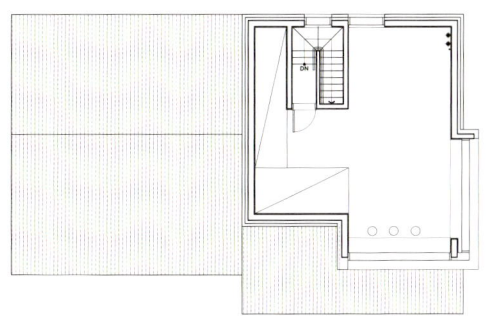

Attic Floor Plan

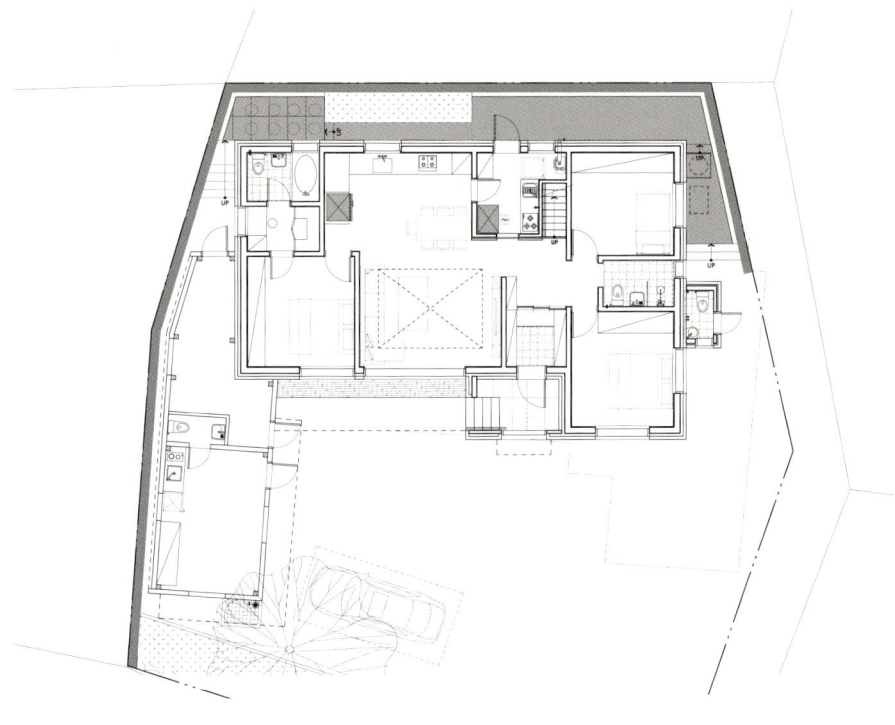

1st Floor Plan

153

YUYUJAJEOG Taean, Korea

admobe architect

Architects admobe architect / Jaehyuk Yi **Location** Taean, Korea **Site Area** 972㎡ **Bldg. Area** 156㎡ **Gross Floor Area** 156㎡ **Bldg. Scale(Floor)** 1FL **Structure** Light-Frame Constructio, Heavy Timber Construction **Max. Height** 6.08m **Exterior Finish** EIFS, Cement Siding(Color Plus) **Construction** Eco-House **Photographer** Changmook Kim

설계 ㈜에이디모베 건축사사무소 / 이재혁 위치 충청남도 태안군 근흥면 수룡리 대지면적 972㎡ 건축면적 156㎡ 연면적 156㎡ 규모 지상1층 구조 경골목구조, 중목구조 최고높이 6.08m 외부마감 EIFS, 시멘트사이딩(칼라플러스) 시공 에코하우스 사진 김창묵

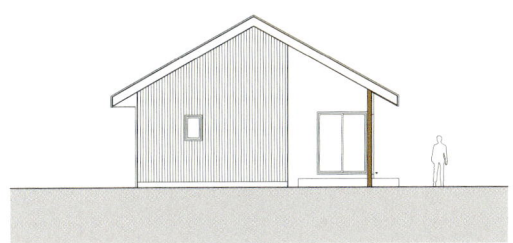

Elevation I

Elevation II

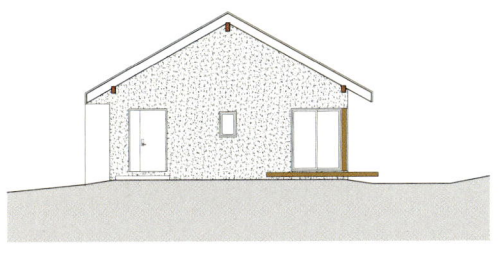

Elevation III

Elevation IV

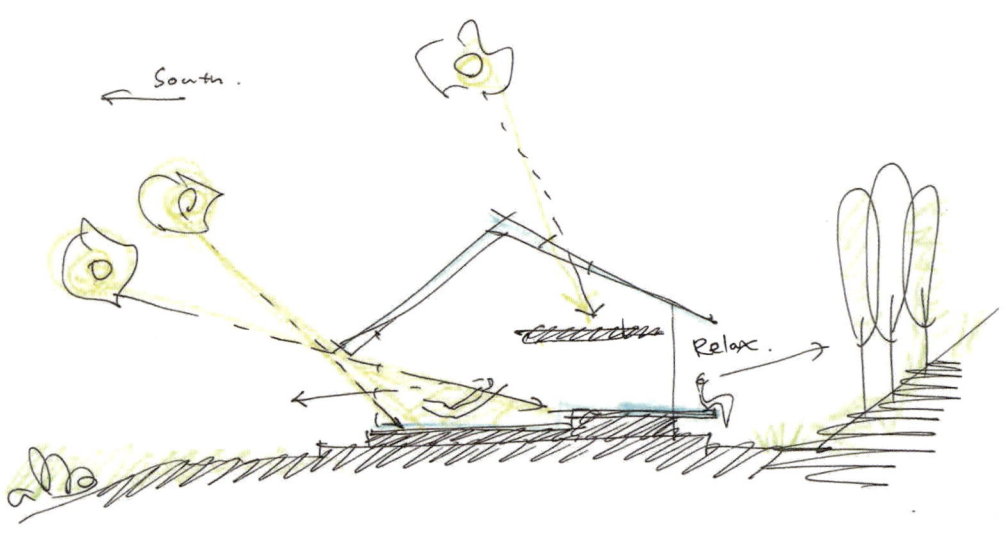

1. T7 ALUZINC(APP' COLOR)
2. WATERPROOF SHEET
3. THK11 OSB ROOF COVER
4. 30×30 SQUARE TIMBER
5. MOISTURE-PERMEABLE AND WATERPROOF RESIN
6. 2×12 RAFTERS @609
7. GLASS WOOL INSULATION
8. THK9.5 GYPSUM BOARD 2PLY
9. APP' FINISH
10. INDIRECT LIGHT(T5 LED)
11. T20 SPRUCE LAMINATED BOARD
12. STRUCTURAL WOOD(120×120, SPRUCE)
13. T16×140 WOOD SIDING
14. PVC SYSTEM WINDOWS
    THK43 LOW-E(SOFT COATING)
    TRIPLE GLASS(5+14+5+14+5)
    "INOUTIC LEGEND SLIDING"
15. THK INSULATION FILLING
16. APP' DECK WOOD
17. INSULATION MATERIAL("A" GRADE EXTRUSION METHOD THERMAL INSULATION PLATE, 70MM)
18. APP' WOOD FLOORING
19. THK83 CEMENT LEVELING MORTAR
    (WIRE MESH #8-150×150)
20. HEATING PIPING
21. PE FILM
22. THK80(50+30) EXTRUSION METHOD THERMAL INSULATION PLATE("A" GRADE)
23. CONCRETE MAT FOUNDATION

1. T7 ALUZINC(지정색)
2. 방수시트
3. THK11 OSB 지붕덮개
4. 30×30 각재
5. 투습방수지
6. 2×12 서까래 @609
7. 글라스울 단열재
8. THK9.5 석고보드 2장
9. 지정마감
10. 간접등(T5 LED)
11. T20 스프러스 집성판
12. 구조목(120×120, 스프러스)
13. T16×140 목재 사이딩
14. PVC시스템창호
    THK43 로이(소프트코팅)
    삼중유리(5+14+5+14+5)
    "INOUTIC LEGEND SLIDING"
15. THK 단열재 충진
16. 지정 데크목
17. 단열재("가"등급 압출법보온판, 70MM)
18. 지정 목재플로링
19. THK83 시멘트 레벨링 몰탈
    (와이어매쉬 #8-150×150)
20. 난방배관
21. PE필름
22. THK80(50+30) 압출법보온판("가"등급)
23. 콘크리트 매트기초

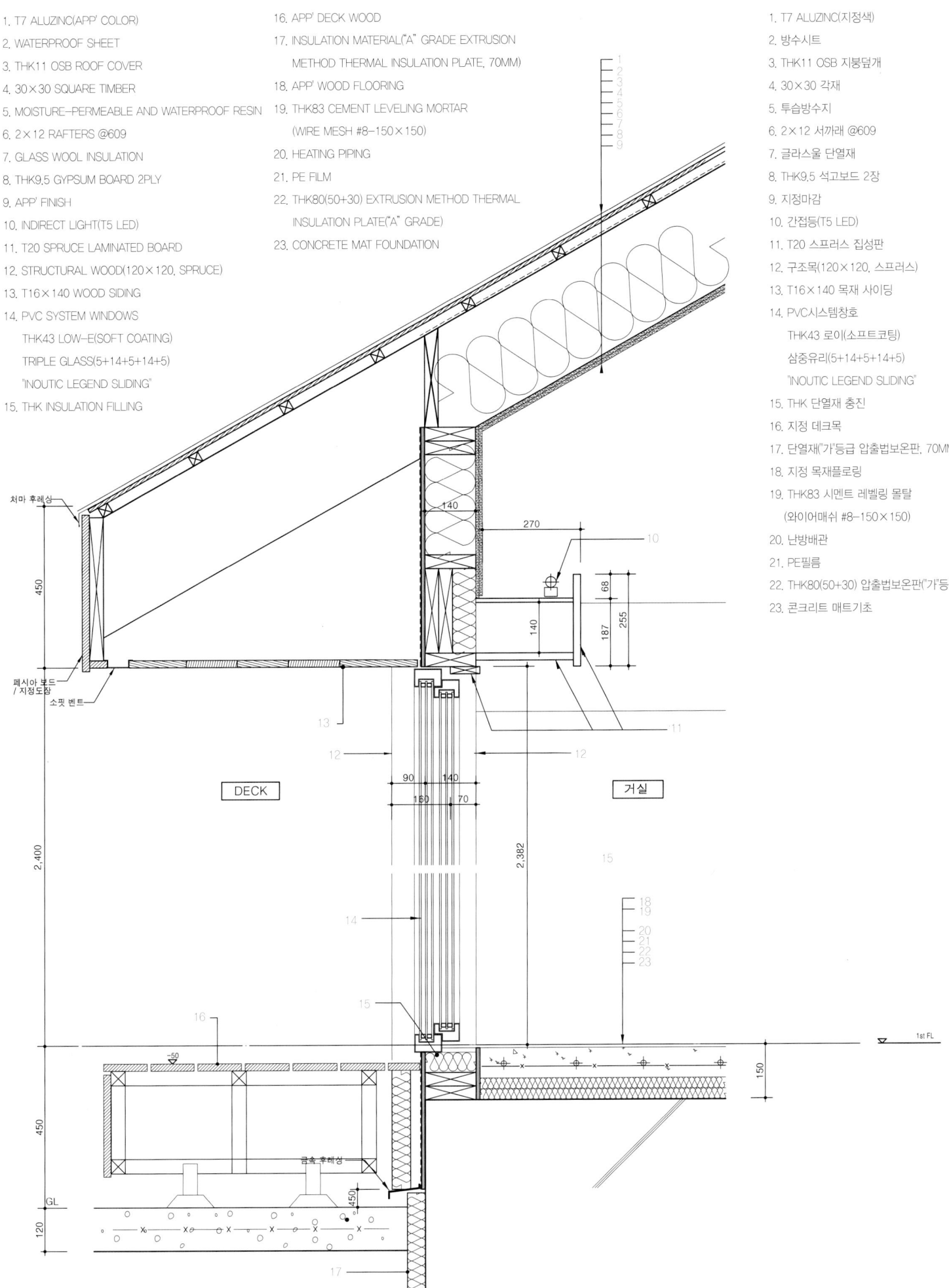

Exterior Wall Section Detail

A retired teacher couple sold their house in Seoul (Apartment in Dongbu Ichon-dong) and built a house to enjoy their old age in Taean, near their hometown. Since I'm a little older, I didn't want to change the furniture and lifestyle I used before, rather than pursuing something new while building a house. It is a work that has become a bit flat for an apartment.

There are about four things I could do as an architect.
1. In order to capture the passion of the client who has studied wood structure construction for a while, a wooden structure is added to the house.
2. Separating the outdoor warehouse, which is important in rural life, from the residential space, and covering the roof in between. We created an outdoor space that can be used even when it snows.
3. A guest room and a study room with a view of the pine trees in the north, and the outside under the roof. It was placed so that the space could be viewed.
4. A vertical and open space was created using an attic and a skylight.

You spent more money than planned to make the roof, but I hope that this space will be used for various purposes as a work space and a party space.

수룡리 단독주택 / 유유자적(悠悠自適)

은퇴하신 선생님 부부가 서울집(동부이촌동 아파트)을 팔고, 고향 가까운 태안으로 내려와서 노후를 즐기기 위한 집을 지었다. 남자 분이 이곳 서산 쪽 출신이신데, 여자 분이 완전 서울 분이라 내려 오시는데 약간의 어려움이 있었다.

연세가 조금 있으시다 보니, 집을 지으면서 새로움을 추구하기 보다는 기존에 쓰시던 가구와 생활패턴을 바꾸고 싶지는 않았다. 해서 약간 아파트 평면이 되어갔던 작업이다.

건축사로서 할 수 있었던 일은 네 가지 정도이다.
1. 한동안 목구조 건축을 공부하셨던 의뢰인의 열정을 담아내기 위해 집 안에 목구조를 조금 노출 시켰다.
2. 전원 생활에서 중요한 옥외 창고를 주거 공간과 분리하면서 그 사이에 지붕을 덮어 비나 눈이 올 때도 사용할 수 있는 옥외 공간을 만들었다.
3. 손님방이자 서재로 쓰일 방으로 북쪽의 소나무를 감상할 수 있도록 하고, 지붕 아래 외부 공간을 바라볼 수 있는 배치를 하였다.
4. 다락방과 천창을 이용해 수직적이고 개방적인 공간을 만들었다.

지붕 만드느라 계획보다 예산도 많이 쓰셨는데 이 공간이 작업 공간이자 파티 공간으로 다양한 용도로 사용되는 공간이 되길 기원해 본다.

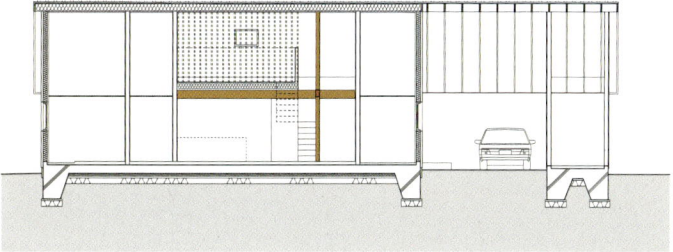

Section A

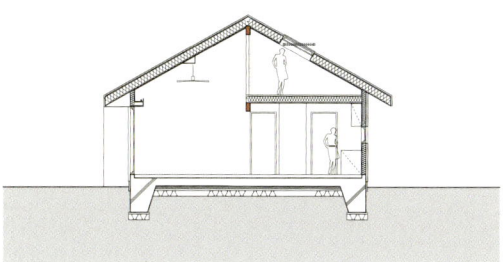

Section B

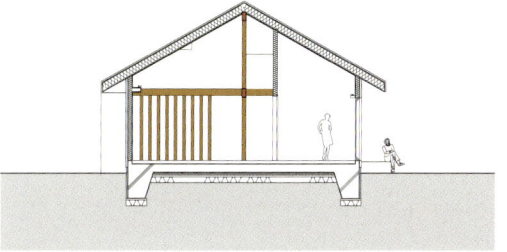

Section C

1. PARKING LOT
   & MULTI-PURPOSE SPACE
2. PORCH
3. BEDROOM
4. LIVING ROOM
5. KITCHEN
6. UTILITY ROOM
7. STUDY
8. DECK
9. WAREHOUSE

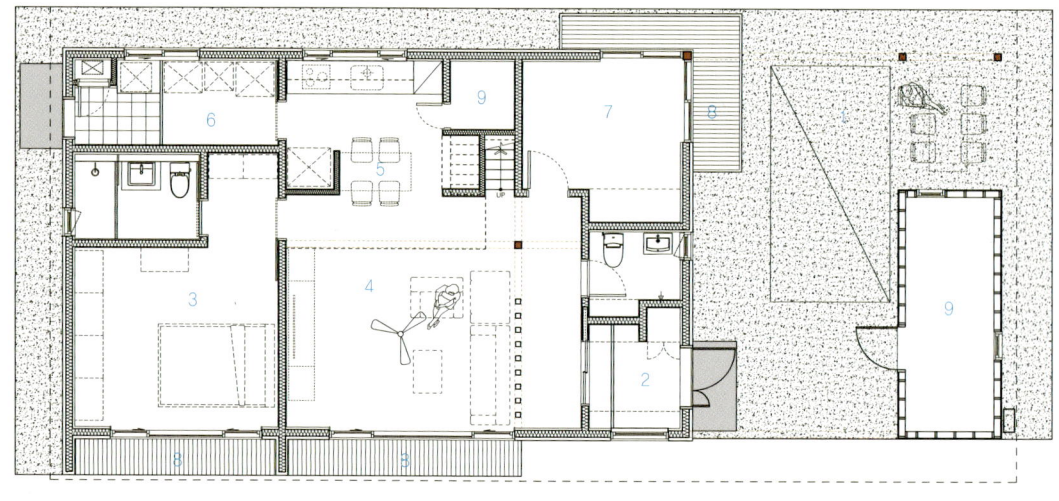

1st Floor Plan

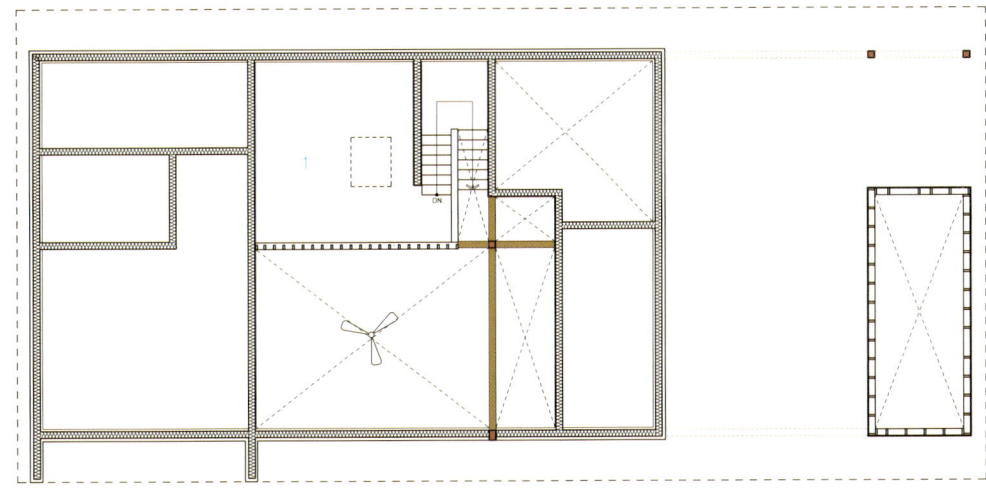

Attic Floor Plan

1. ATTIC

Boryeong House Boryeong, Korea

WELLHOUSE

Architects WELLHOUSE / Jeguen Myung **Location** Boryeong, Korea **Site Area** 985㎡ **Bldg. Area** 172.78㎡ **Gross Floor Area** 247.38㎡ **Bldg. Scale(Floor)** 2FL **Structure** Reinforced Concrete **Photographer** Younggi Min

설계 웰하우스종합건축사사무소 / 명제근 **위치** 충청남도 보령시 청소면 성연리 **대지면적** 985㎡ **건축면적** 172.78㎡ **연면적** 247.38㎡ **규모** 지상2층 **구조** 철근콘크리트 **사진** 민영기

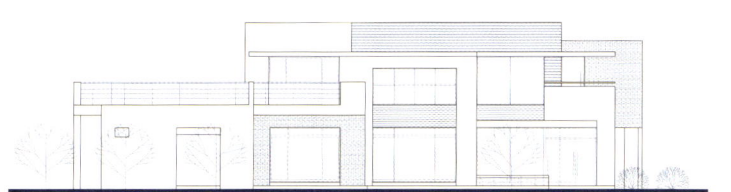

Front Elevation

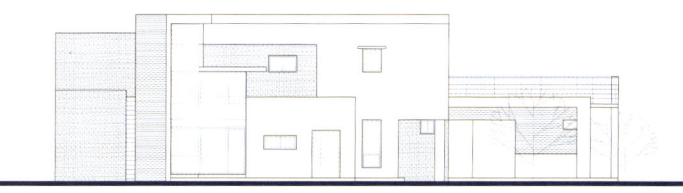

Rear Elevation

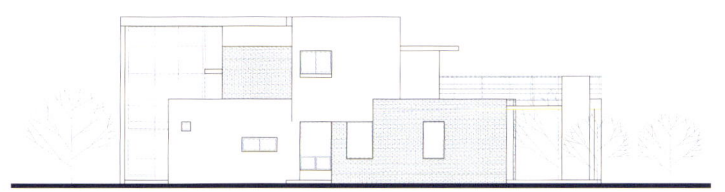

Left Elevation

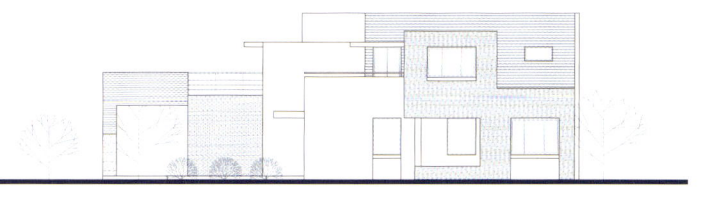

Right Elevation

It is located at a suitable height to view Ansan(front Mt.) in front of the site. It is an environment where you can see a beautiful sunset in harmony with Oseosan Mountain, which has a beautiful natural topography, and Josan(distant Mt.) Mountain in the far southwest, in harmony with the surrounding mountains. Considering this site environment, the axis of the house was planned to face southeast, and the view and communication from the outside space are made through the veranda and bridge for the observation on the second floor. The window of the stairwell connecting the vertical movement becomes another sub-view element of the house that contains the beautiful natural environment of Mt. Oseo.

보령 주택

대지의 앞쪽으로 안산이 바라보기에 적당한 높이에 위치하고 있다. 뒤쪽으로는 자연지세가 아름다운 오서산과 남서쪽 멀리 조산이, 주변 산세와 어우러져 아름다운 석양을 볼 수 있는 환경이다. 이러한 대지 환경을 고려하여 주택의 축은 남동향으로 계획하였고, 2층에 전망용 베란다와 브릿지를 통해 조망과 외부공간에서의 소통이 이루어진다. 수직 동선을 연결하는 계단실의 창은 오서산의 아름다운 자연환경을 담는 집의 또 다른 차경 요소가 된다.

1. BED ROOM
2. DRESS ROOM
3. BATH ROOM
4. LIVING ROOM
5. KITCHEN
6. BOILER ROOM
7. DINING ROOM
8. FOYER
9. FRONT ROOM

1st Floor Plan

1. FAMILY ROOM
2. BED ROOM
3. BATH ROOM
4. TERRACE
5. DECK

2nd Floor Plan

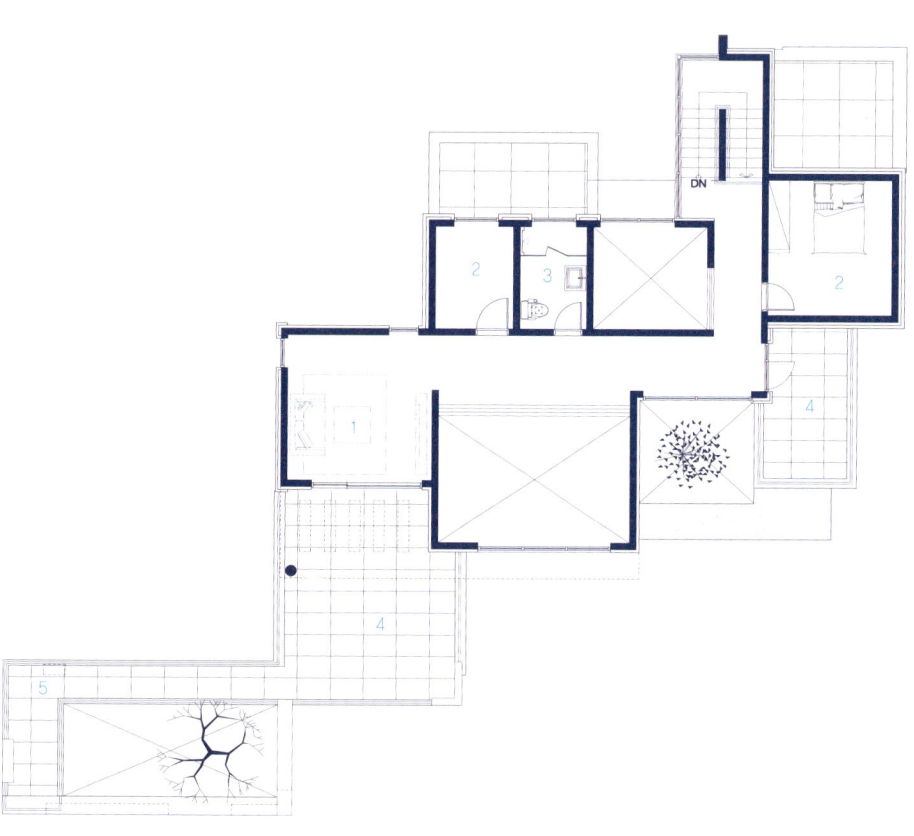

Pyeongdamjae Jangseong, Korea

Office for Appropriate Architecture

Architects Office for Appropriate Architecture / Jooyoun Yoon **Project Team** Jiwoo Jeong, Seungwon Lee **Location** Jangseong, Korea **Site Area** 613m² **Bldg. Area** 199.9m² **Gross Floor Area** 183.2m² **Bldg. Scale(Floor)** 1FL, Attic **Structure** Wood Structure **Max. Heigh** 5.3m **Exteriror Finish** Brick **Construction** Edenheim **Photographer** Wonseok Lee

설계 적정건축 / 윤주연 설계참여자 정지우, 이승원 위치 전남 장성군 진원면 진원리 대지면적 613m² 건축면적 199.9m² 연면적 183.2m² 규모 지상1층, 다락 구조 목구조 최고높이 5.3m 외부마감 벽돌 시공 이든하임 사진 이원석

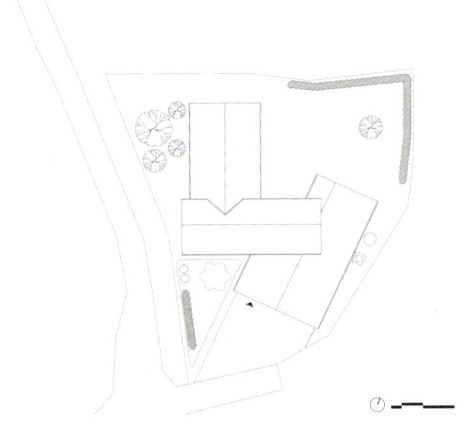

Site Plan

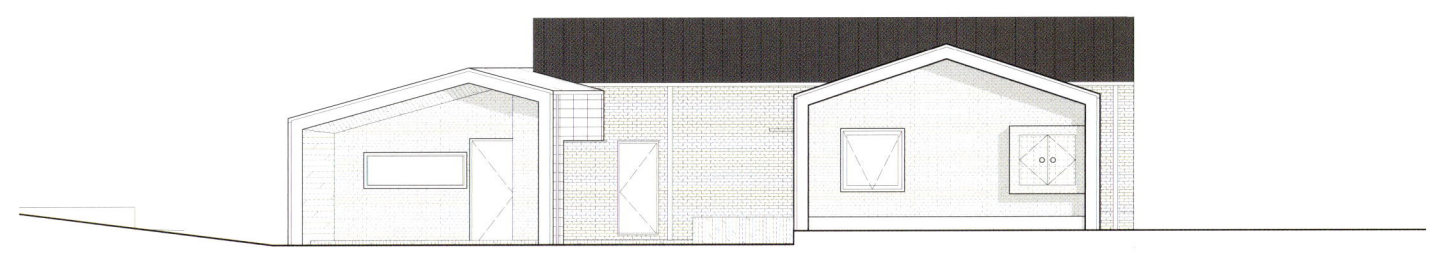

Elevation I

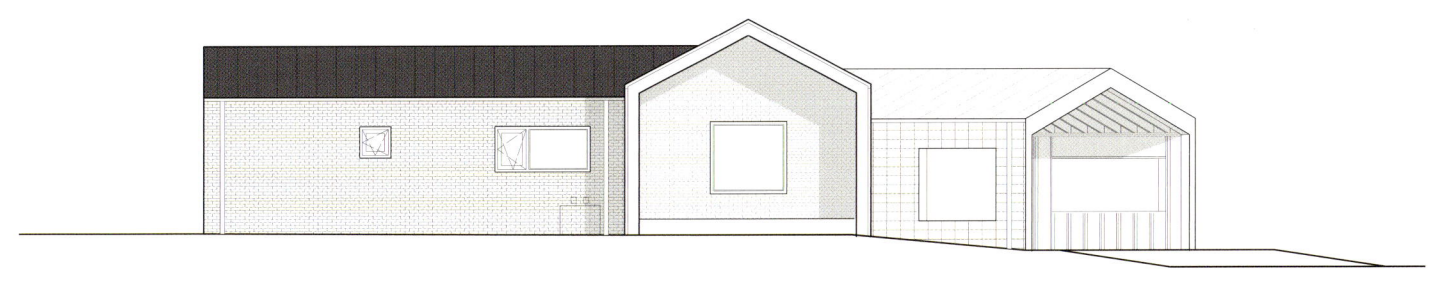

Elevation II

'Pyeongdamjae' is a house for a professor and a middle-aged couple who are ex-French chefs preparing for their retirement in 10 years. The owner requested the following space.

1. Two spaces such as the main house and the Sarangchae - A living room of about 37 pyeong and a guesthouse where guests will stay when they come.
2. Not a house with green grass, but a courtyard with privacy - A house with a backyard in a courtyard where no one can see you even if you wear shabby clothes
3. A space where a hanok and a Western-style house harmonize - I like the floor, but what I needed was a porch, a patio, a backyard, and a fence, and it required a delicate movement line like a Japanese house.
4. I wanted a single floor because I have bad knees, but I also wanted a space that was partially on the second floor.

I visited the site and thought that it was important to match well with the neighborhood while placing importance on privacy due to the couple's personality. I wanted a yard and a fence, but considering the budget range, I divided the yards with different characteristics according to the layout of the house. These are the American-style porch (front yard) where you can chat comfortably with your neighbors, the backyard, a space for families, and the planting yard.

Among the requirements for the whole house, the part for guests and annexes was large. The space for guests was important because of the relationship I made while staying abroad for a long time, but it was important to find a suitable space that would not harm the space of the couple who always use the house due to the low frequency of use. We put a small duplex space in the study room so that guests can rest comfortably in the study room like an annex. The guest space and family space meet at the dining table. It is a shape where the family zone extending to the north and the guest zone extending to the east are intertwined with the dining shared by both family and guests at the center.

At the time of design, two points of design were obtained by observing the house where the owner resides. First, the kitchen, my wife's space, had a lot of kitchen living. I needed a more spacious kitchen than others, and I wanted the way I cook to be the most important point in the house. It's a chef's cooking space, not buokte. There is a dining table for 6 in front of it, so it can be used generously, but it is also intended for tea time and light meals for couples to enjoy at a small counter in front of the kitchen.

Another is my husband's space. My husband, whom I introduced as 'our gardener' from the first meeting, is an ecologist and has decided to take charge of the space outside the house. But, like most fathers, there is no room for him in the house. So he roamed around the house and saw what he had to work on, which got him thinking about his husband's space. He has a small room in the corner of the living room where he can do light paperwork while feeling connected to his family. Although it is a small room with a width of 1.8 meters, there is a spacious desk made with a wooden top made on site. This room has a double sliding door, and when the door is closed, it becomes a small room that can sleep two people. It is a small family room where the daughters who are out of the country can rest when they return.

The single-story house creates an active relationship with the external space as it is widely spread on the site. Windows of various heights and shapes are provided to provide the most consistent atmosphere and view for each space. In the dining space, which is the center of the house, there is a large window seat on the west side so that the idyllic scenery of rice fields can be well captured, while in the couple's bedroom there is a low and wide window through which the morning sunlight softly enters. In addition to the use of space, Jangjimun creates a calm hanok-like atmosphere that does not require a separate awning facility.

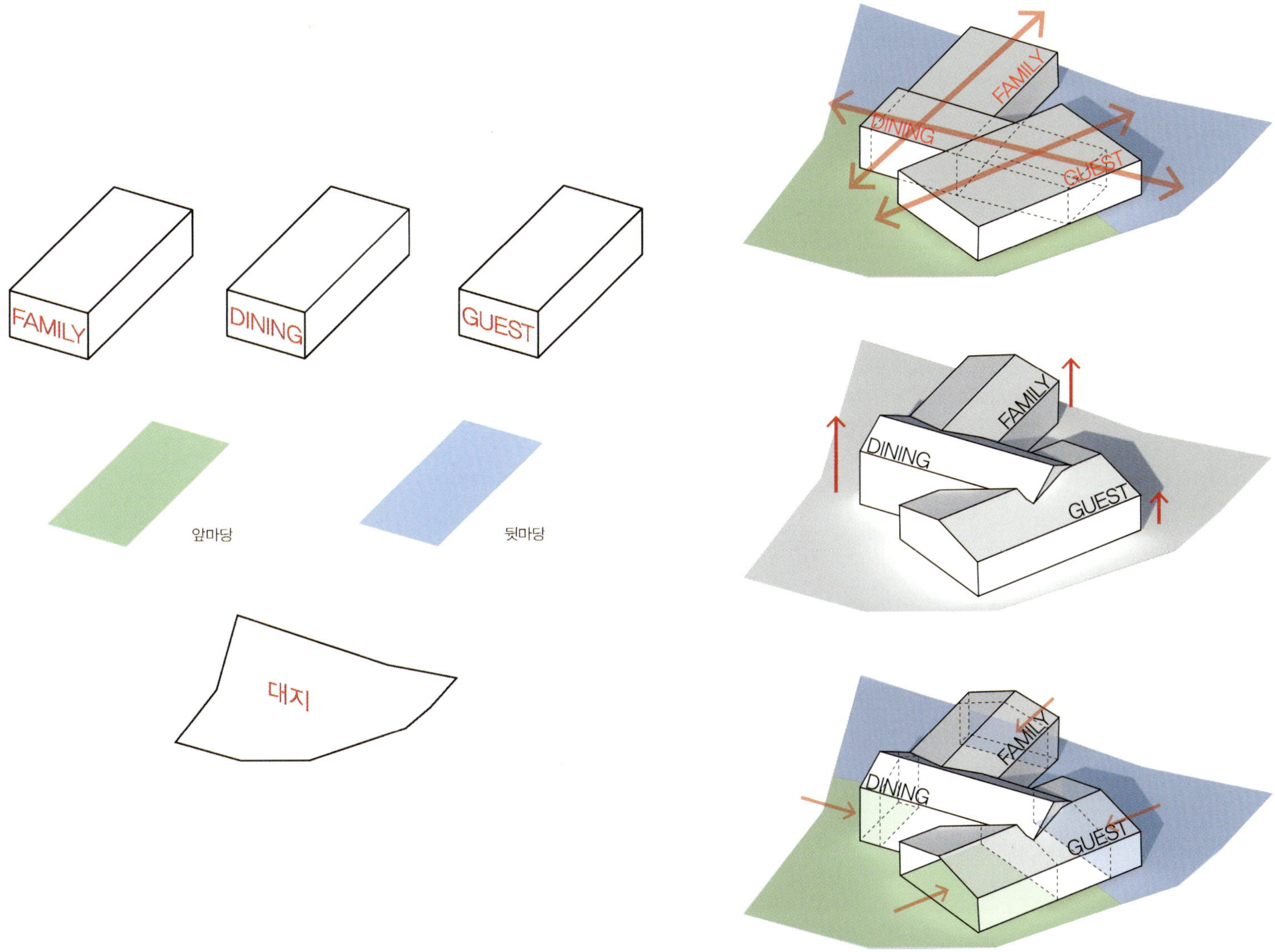
FAMILY
DINING
GUEST
앞마당
뒷마당
대지
FAMILY
DINING
GUEST
FAMILY
DINING
GUEST
FAMILY
DINING
GUEST

Diagram

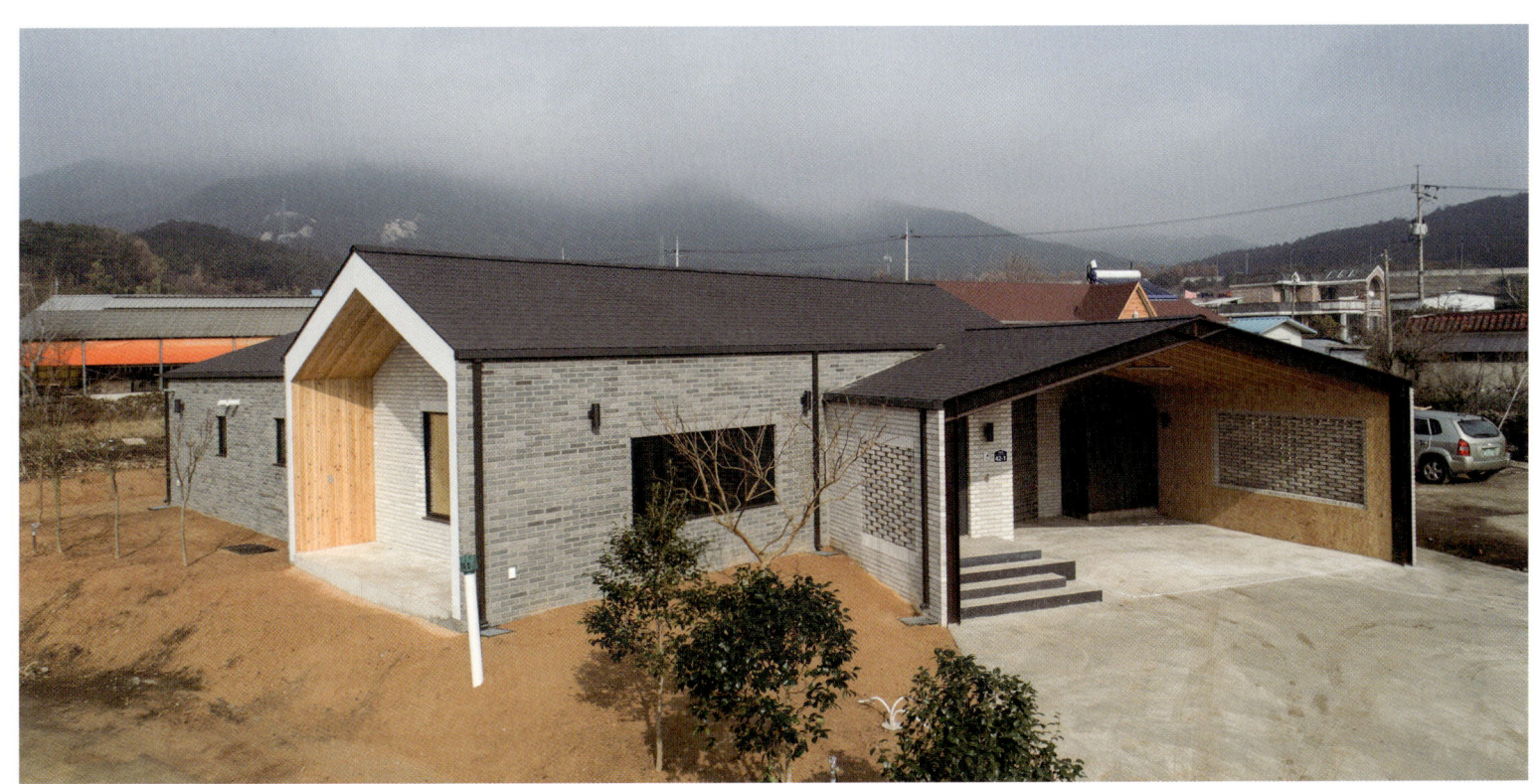

평담재

'평담재'는 10년 후 은퇴생활을 미리 준비하는 교수님과 전직 프랑스 요리사인 중년 부부를 위한 집이다. 건축주는 아래와 같은 공간을 요구하였다.

1. 안채와 사랑채 같은 두 개의 공간 – 37평 정도의 살림집과 손님이 오면 묵으실 게스트하우스겸 서재.
2. 파란 잔디가 깔린 집이 아니라 프라이버시가 확보된 안마당 – 허름하게 입고 다녀도 볼 사람없는 안마당 뒷마당이 있는 집.
3. 한옥과 양옥이 어우러진 공간 – 툇마루가 좋지만 필요한건 포치와 파티오, 백야드 그리고 담장이었으며, 일본주택 같은 섬세한 동선을 필요로 하였다.
4. 무릎이 좋지 않아 단층을 원했지만 일부분 2층인 공간도 원하였다.

현장을 답사하고 배치에는 부부의 성격상 프라이버시를 중요시하면서도 동네와 잘 어울릴 것을 중요하게 보았다. 마당과 담장을 원하였는데, 예산범위를 생각하여 집의 배치로 성격이 다른 마당들을 나누었다. 동네 이웃과 편히 담소를 나눌수 있는 미국식 포치(앞마당), 가족만의 공간인 백야드, 식재 마당들이 그것이다.

집 전체에 대한 요구사항 중 손님과 별채에 대한 부분이 컸다. 오랫동안 해외에 체류하며 맺은 인연으로 손님 치를 일이 많아 손님을 위한 공간이 중요했지만 평상시에 활용빈도가 낮아 늘상 집을 사용하는 부부의 공간을 해치지 않는 적정선을 찾는 것이 중요했다. 서재에 작은 복층 공간을 넣어 손님이 서재를 별채처럼 내어 편히 쉬실 수 있게 하였다.

손님공간과 가족공간은 식탁에서 만난다. 가족과 손님 모두 공동으로 사용하는 다이닝을 중심에 두고 북쪽으로 확장하는 패밀리존과 동쪽으로 확장하는 게스트존이 엮이는 형상이다. 설계 당시 건축주가 거주하는 집을 관찰하여 두가지 설계의 포인트를 얻었다. 먼저 아내의 공간인 주방은 주방살림이 굉장히 많았다. 남들보다 넉넉한 주방이 필요하였고, 요리를 하는 모습이 집의 포인트가 되면 좋겠다 싶었다. 부엌떼기가 아닌 쉐프의 요리공간으로 말이다. 그 앞에 6인용 식탁이 있어 넉넉하게 쓸수도 있지만, 주방 앞에 작은 카운터에서 부부가 즐기는 티타임과 가벼운 식사도 의도된 것이다.

또 하나는 남편의 공간이다. 처음 만날때부터 '우리집 정원사'라고 소개하신 남편분은 생태학자이며 집 외부공간을 책임지기로 하였다. 하지만 일반적인 아빠들이 그러하듯이 집안에는 그의 공간이 없다. 그래서 집안 여기저기를 돌아다니며 작업해야 하는 모습을 보고 남편의 공간에 대해 생각하게 되었다. 가족과 유대를 느끼면서도 가벼운 서류 작업을 할 수 있는 작은 방을 거실 한켠에 마련했다. 1.8m 폭의 작은 방이지만 현장에서 제작한 목재상판을 얹어 만든 넓직한 책상이 생겼다. 이방은 양개 미닫이 문으로 되어 있어 문을 닫으면 2명이 잘 수 있는 작은방이 된다. 타지에 나가있는 딸들이 돌아오면 쉴 수 있는 작은 패밀리룸이다.

단층 집은 대지에 넓게 퍼져있음으로 외부공간과 적극적인 관계를 만든다. 각 공간마다 가장 적합한 분위기와 뷰를 선사할 수 있도록 다양한 높이와 형태의 창문을 내었다. 집의 중심인 다이닝공간에는 서쪽에 위치한 논의 목가적인 풍경이 잘 담길 수 있게 큰 윈도위 시트가 있다면, 부부침실에는 아침 햇살이 은은하게 들어오는 낮고 넓은 창이 있다. 장지문은 공간의 활용과 더불어 별도의 차양시설이 필요없는 차분한 한옥 같은 분위기를 만들기도 한다.

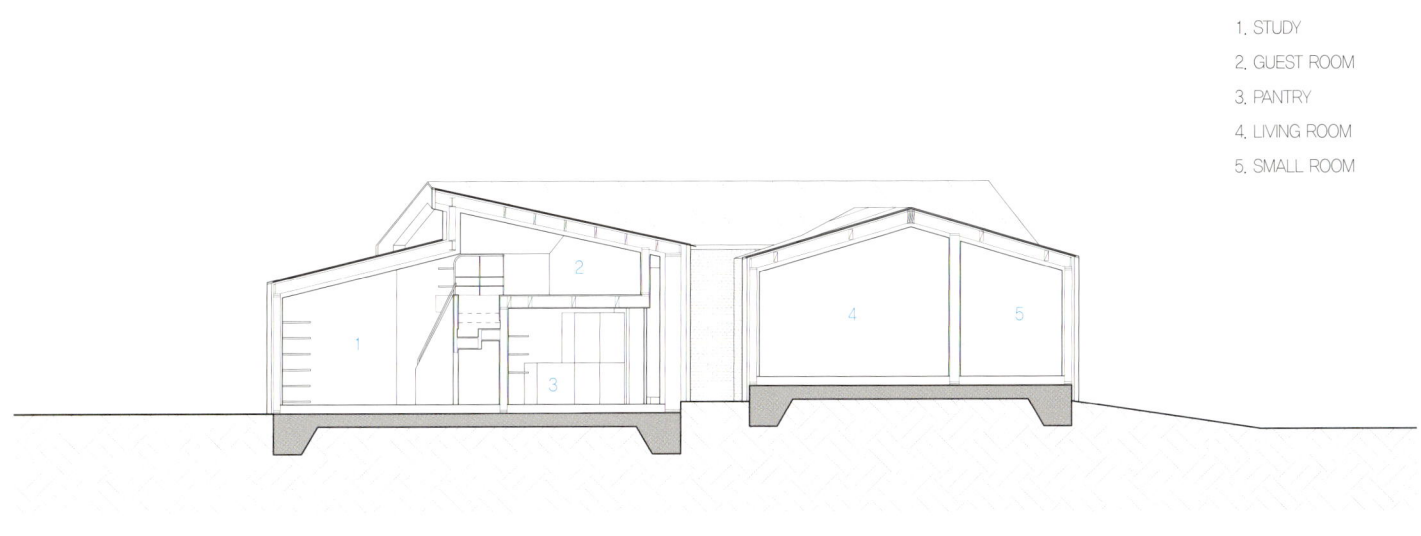

Section

1. STUDY
2. GUEST ROOM
3. PANTRY
4. LIVING ROOM
5. SMALL ROOM

1. WOOD SIDING / OIL STAINLESS
2. APP' OLD BRICK TILE
   MESH PLASTERING / 50T EPS
   TYVEK SUPRO / MOISTURE-PERMEABLE WATERPROOF PAPER
   T11 OSB STRUCTURAL PLYWOOD
   R21 FIBERGLASS INSULATION
   2"×6" STRUCTURAL MATERIAL
   DOPONT AIR GUARD SD5 / AIRTIGHT PAPER
   T12 GYPSUM BOARD 1PLY
   APP' LAMINATED WALLPAPER
3. ASPHALT SHINGLE
   T2.0 WATERPROOF SHEET
   T11 OSB STRUCTURAL PLYWOOD
   2"×2" PRESERVED WOOD
   TYVEK SUPRO / MOISTURE-PERMEABLE WATERPROOF PAPER
   RAFTER 2"×10'(38×235) / R-32 FIBERGLASS INSULATION
   DOPONT AIR GUARD / AIRTIGHT PAPER
   2"×4" SQUARE WOOD(38×89) / SECURE 100 OF FACILITY SPACE
   T12 GYPSUM BOARD 1PLY
   APP' LAMINATED WALLPAPER
4. APP' OLD BRICK TILE
   MESH PLASTERING / 50T EPS
   TYVEK SUPRO / MOISTURE-PERMEABLE WATERPROOF PAPER
   T11 OSB STRUCTURAL PLYWOOD
   R21 FIBERGLASS INSULATION
   2"×6" STRUCTURAL MATERIAL
   DOPONT AIR GUARD SD5 / AIRTIGHT PAPER
   T12 GYPSUM BOARD 1PLY
   APP' PAINT

1. 우드사이딩 / 오일스텐
2. 지정 고벽돌타일
   매쉬미장 / 50T EPS
   TYVEK SUPRO / 투습방수지
   T11 OSB 구조용합판
   R21 유리섬유단열재
   2"×6" 구조재
   DOPONT AIR GUARD SD5 / 기밀방습지
   T12 석고보드 1겹
   지정 합지벽지
3. 아스팔트 쉼글
   T2.0 방수쉬트
   T11 OSB 구조용합판
   2"×2" 방부목상
   TYVEK SUPRO / 투습방수지
   서까래 2"×10'(38×235) / R-32 유리섬유단열재
   DOPONT AIR GUARD / 기밀방습지
   2"×4" 각재(38×89) / 설비공간 100확보
   T12 석고보드 1겹
   지정 합지벽지
4. 지정 고벽돌타일
   매쉬미장 / 50T EPS
   TYVEK SUPRO / 투습방수지
   T11 OSB 구조용합판
   R21 유리섬유단열재
   2"×6" 구조재
   DOPONT AIR GUARD SD5 / 기밀방습지
   T12 석고보드 1겹
   지정도장

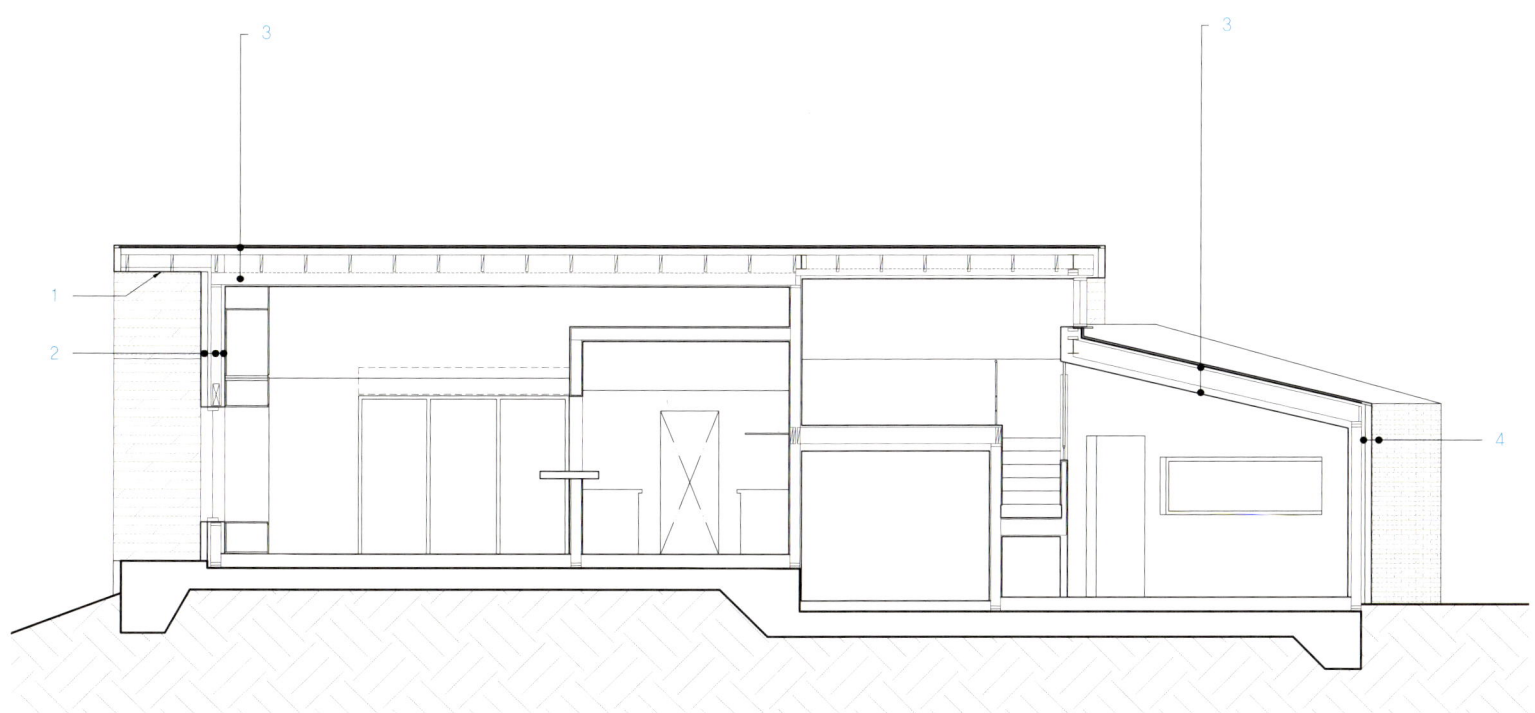

Section Detail

Axonometric

1. PORCH	8. STUDY
2. PARKING LOT	9. LIVING ROOM
3. DINING	10. SMALL ROOM
4. KITCHEN	11. BEDROOM
5. PANTRY	12. BATHROOM
6. SECONDARY KITCHEN	13. DRESSING ROOM
7. TOILET	14. GUEST ROOM

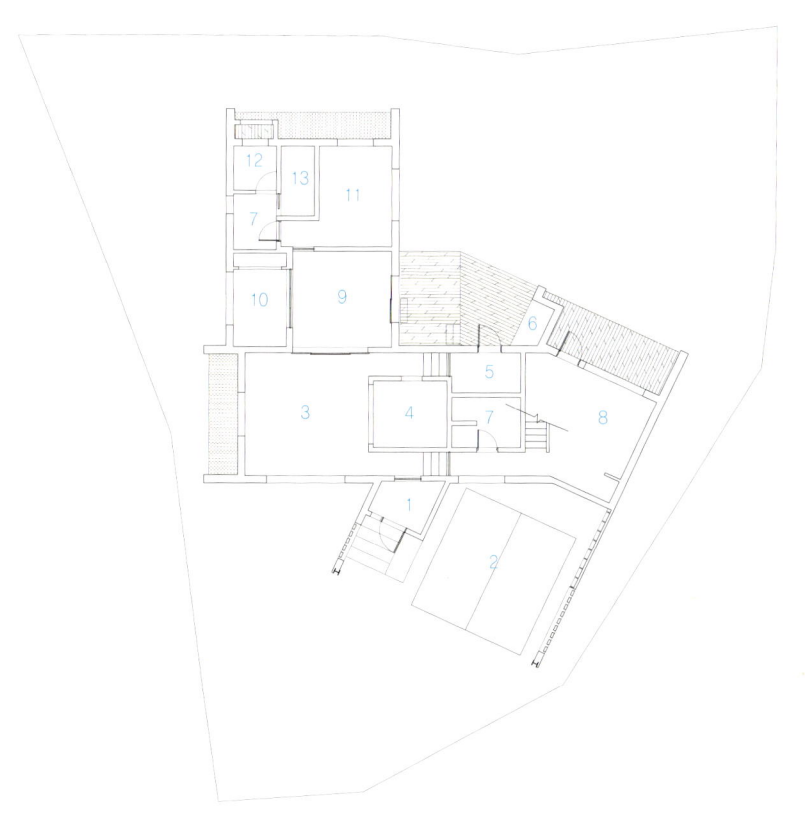

1st Floor Plan

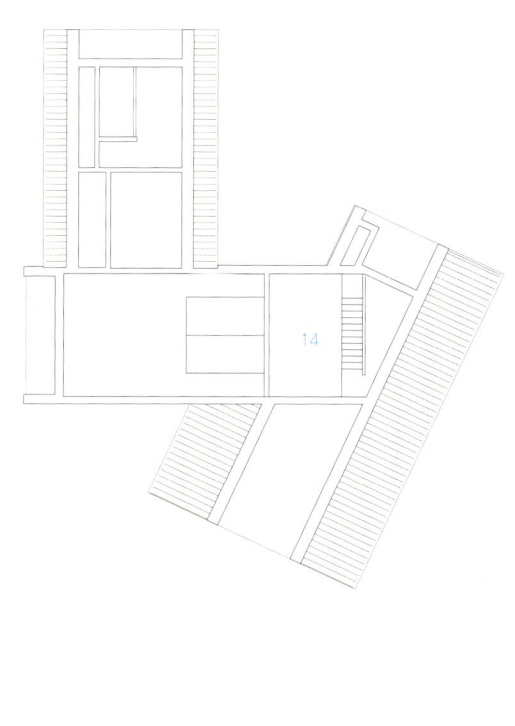

Attic Floor Plan

Mok-dong House Gwangju, Korea

JDArchitects

Architects JDArchitects **Project Team** Jeongdo Choi **Location** Gwangju, Korea **Use** House **Site Area** 496m² **Bldg. Area** 155.11m² **Gross Floor Area** 276.21m² **Bldg. Scale(Floor)** 2FL, B1 **Construction** Wonil Kim **Client** Taeja Kim **Photographer** Namseon Lee

설계 제이디에이건축사사무소 설계참여자 최정도 위치 경기도 광주시 목동 용도 하우스 대지면적 496㎡ 건축면적 155.11㎡ 연면적 276.21㎡ 규모 지상2층, 지하1층 시공 김원일 발주처 김태자 사진 엘투아카이브 이남선

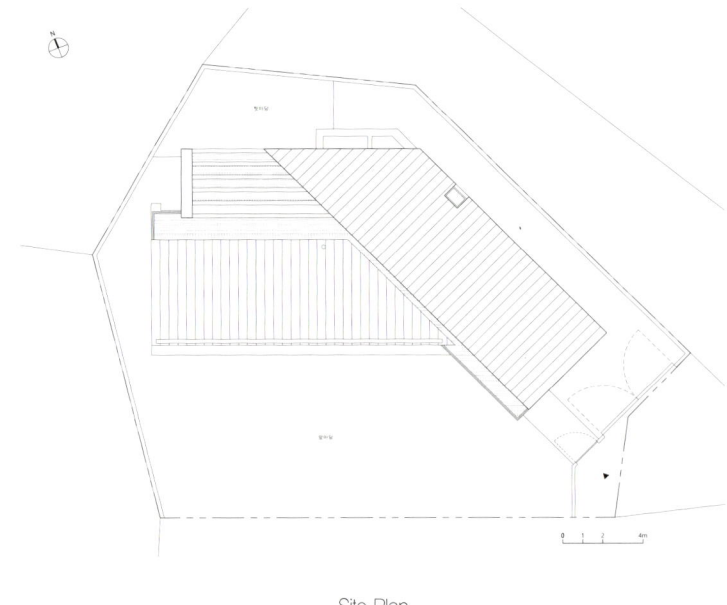

Site Plan

Front Elevation

Right Elevation

Rear Elevation

Left Elevation

The site, located in Mokdong, Gwangju-si, Gyeonggi-do, is surrounded by low mountains and has a refreshing view and a cozy atmosphere. As I looked around the land located at the entrance of the rural housing complex, which was almost completely developed, I imagined an elegant house that fits well with the natural environment while giving fresh energy to the neighborhood.

The shape of the earth was difficult to seat the house. It is straight to the south, but it has an atypical shape to the north, so it was not easy to prepare the spacious front yard that the client expects. Eventually, the masses on the first and second floors were alternately arranged according to the shape of the site, and a wall and slabs with a zinc finish were arranged to enclose the two masses. The mixed mass was well suited to the needs of the owner, who emphasized the privacy of each bedroom. The naturally created piloti space was used as a parking lot, and a terrace with a wooden deck was created on the second floor. The eaves created by protruding the slabs surrounding the mass created a comfortable indoor atmosphere while controlling the amount of sunlight. From the living room on the south side, you can see the comfortable ridge of Mt. Munhyung, passing the field cultivated by the client, and the kitchen and dining room on the north side, you can see the mountain peak with the view of the village below. On the second floor deck, which is shaded with louvers, you can enjoy the nature with the mountains open in all directions. The client was satisfied with being able to sit in the bathtub in the master bedroom on the second floor and see Mt. Neungan in the distance.

The Mokdong detached house was a house planned by a middle-aged client to live a long life. The exterior material also had to be able to withstand the weight of time. The main material surrounding the house is bright titanium zinc, which reduces the burden of maintenance. In order to prevent contamination of the external insulation system on the lower floor, attention was paid to the details of the flashing around the window.

Any architect will remember the project they met for the first time on their own Mokdong House was such a project. The client waited for the plan with a heart of trust, and thanks to this, he was able to realize his deep concerns through architecture. There was even the fortune of meeting a builder who silently endured the rough drawings and the fickle scene. After all, it was a project that once again realized that good architecture comes from the relationship between the client, the architect, and the constructor.

1. THK0.7 ZINC STANDING SEAM @430	1. THK0.7 징크 돌출이음 @430
2. SILICON SEALANT	2. 실리콘실란드
3. THK0.8 ALUMINUM FLASHING	3. THK0.8 알미늄후레싱
4. URETHAN FOAM	4. 우레탄폼 사춤
5. BACKUP MATERIAL / SILICON CAULKING	5. 백업재 / 실리콘 코킹
6. 12T PLYWOOD	6. 12T 내수합판
7. GIWS WATERPROOF SHEET	7. GIWS 방수시트
8. THK0.7 ZINC	8. THK0.7 징크
9. 2.0T ASPHALT SHEET	9. 2.0T 아스팔트시트
10. ST'L PIPE 40×40×1.6T	10. 각파이프 40×40×1.6T
11. THK80 INSULATOR / T0.05 P.E FILM 2LAYERS	11. THK80 단열재 / T0.05 PE필름 2겹
12. THK20 PROTECTION MORTAR	12. THK20 보호몰탈
13. URETHAN FILM WATERPROOFING	13. 우레탄 도막방수
14. CON'C TROWEL FINISH	14. 콘크리트쇠흙손마감
15. DOUGLAS-FIR CEILING BOX	15. 미송천장틀
16. T12 NATURAL WOOD FINISH	16. T12천연목재마감

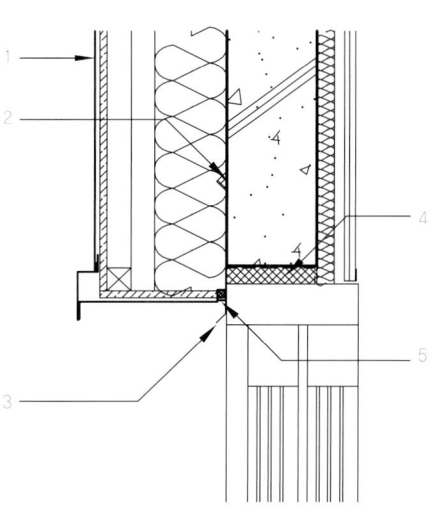

Flashing Section Detail A

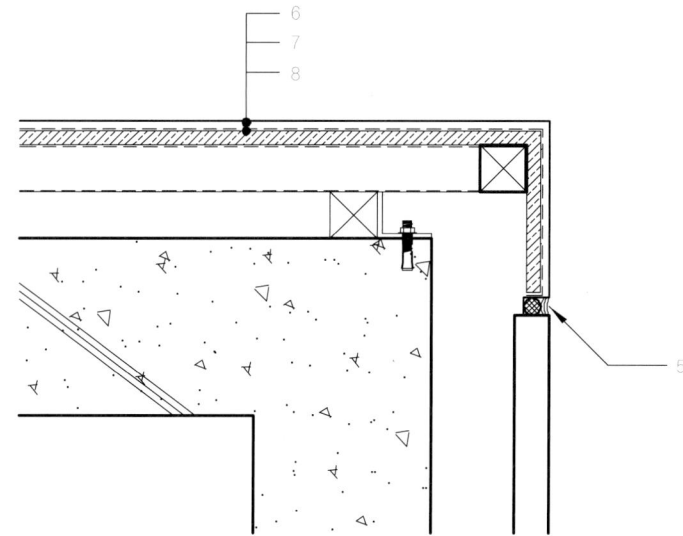

Flashing Section Detail B

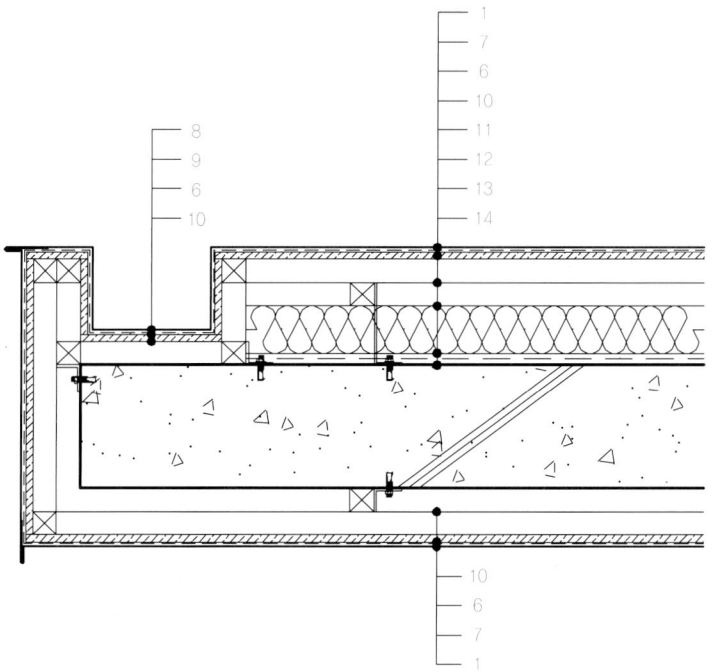

Zinc Panel Section Detail A

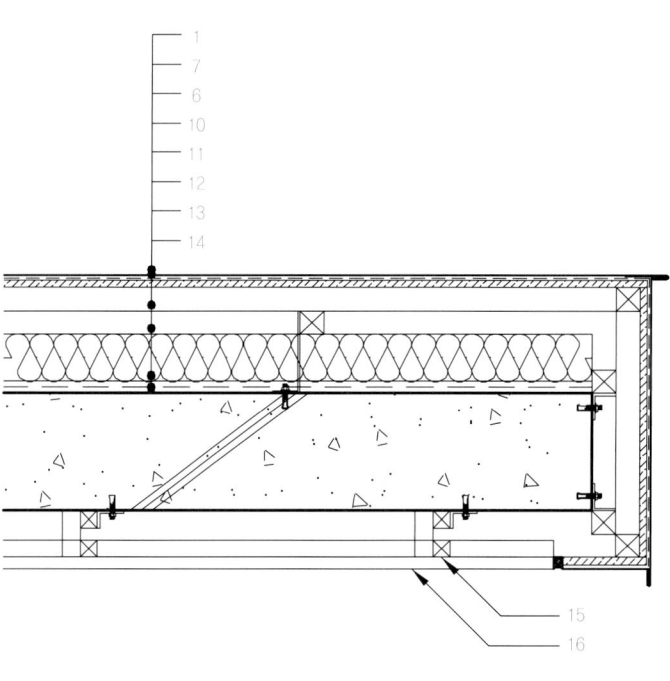

Zinc Panel Section Detail B

목동주택

경기도 광주시 목동에 위치한 대지는 나지막한 산으로 둘러쌓여 탁트인 전망과 함께 포근한 분위기가 흐르고 있었다. 개발이 거의 완료된 전원주택 단지의 초입에 위치한 땅을 둘러보면서 동네에 신선한 기운을 주면서도 자연환경에 잘 스며드는 단아한 주택을 상상했다.

대지의 형상은 집을 앉히기 까다로웠다. 남쪽으로는 반듯하나 북쪽으로 비정형의 모양을 가지고 있어 건축주가 기대하는 넓직한 앞마당을 마련하기 쉽지 않았다. 결국 대지의 형태에 따라 1층과 2층의 매스를 엇갈려 배치하고 징크마감의 벽과 슬라브가 두 매스를 감싸 안는 안을 마련했다. 엇갈린 매스는 각 침실의 프라이버시를 강조한 건축주의 요구에도 잘 맞았다. 자연히 만들어진 필로티 공간을 주차장으로 사용하고 2층에는 목재데크로 마감된 테라스가 생성되었다. 매스를 감싸는 슬라브를 돌출시켜 생성된 처마는 일사를 조절하면서도 편안한 실내 분위기를 만들어주었다.

남측의 거실에서는 건축주가 경작하는 밭은 지나 문형산의 편안한 능선이 보이고 북측의 주방과 식당에서는 아랫동네 전경과 함께 너락봉을 조망할 수 있다. 루버로 그늘을 만든 2층 데크에서는 사방으로 트인 산세와 함께 자연을 만끽할 수 있도록 했다. 건축주는 2층의 안방 욕조에 앉아 멀리 능안산을 바라볼 수 있는 것에 만족해했다.

목동 단독주택은 중년의 건축주가 오래도록 살기위해 계획한 집이었다. 외장재도 시간의 무게를 견딜 수 있어야 했다. 집을 감싸안는 주 재료는 밝은 티타늄징크를 써서 유지관리의 부담을 덜었다. 저층부 외단열시스템의 오염을 막기위해 창호주위 후레싱 디테일에 신경을 썼다.

건축사라면 누구나 독립하여 처음으로 만난 프로젝트를 각별히 기억할 것이다. 목동 단독주택이 그런 프로젝트였다. 건축주는 신뢰의 마음으로 계획안을 기다려 주었고 덕분에 깊은 고민을 건축으로 실현시킬 수 있었다. 거친 도면과 변덕스러운 현장을 묵묵히 버텨준 시공자를 만나는 행운까지 있었다. 결국 좋은 건축은 건축주, 건축사, 시공자의 관계 속에서 나온다는 것을 다시 한번 실감한 프로젝트였다.

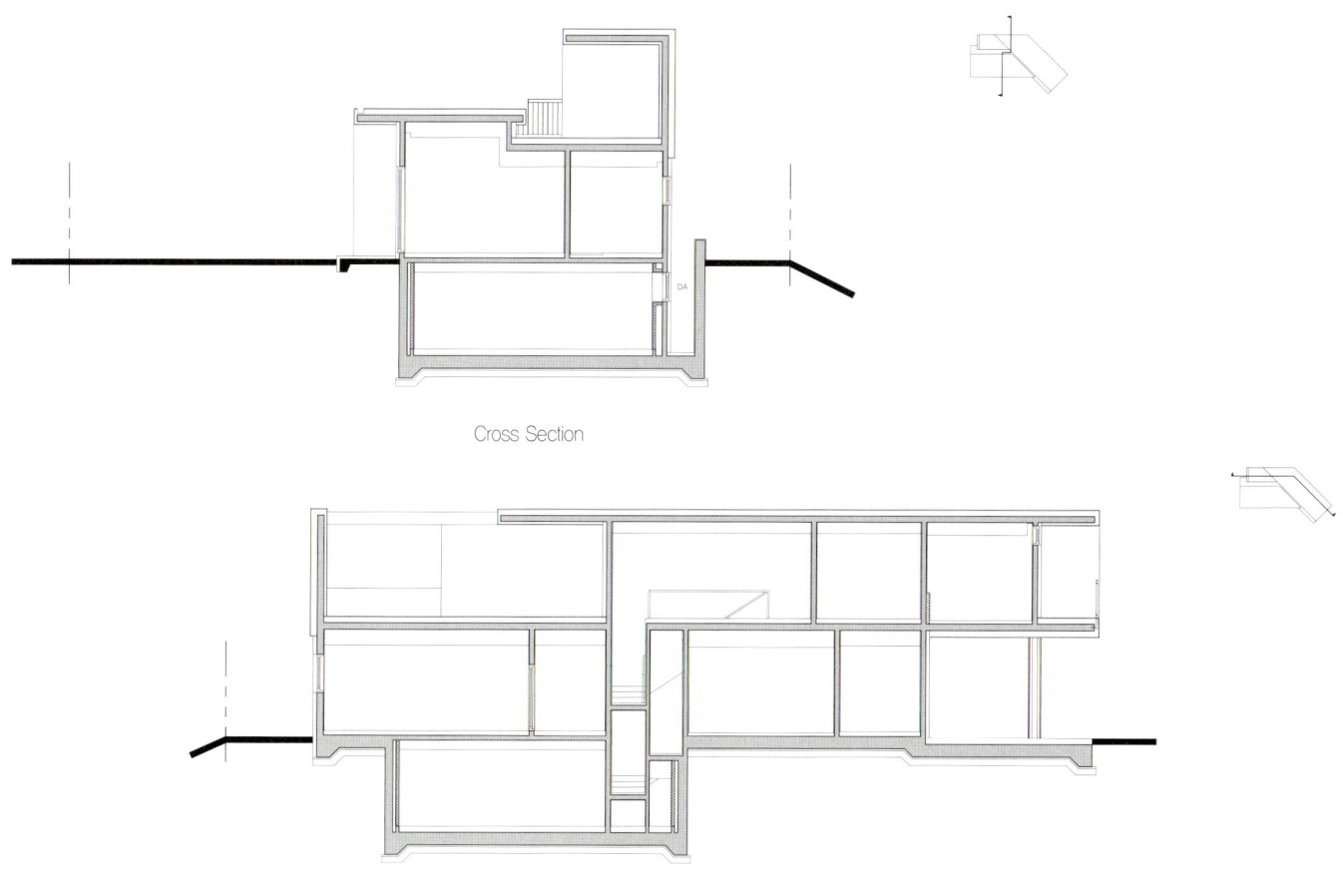

Cross Section

Longitudinal Section

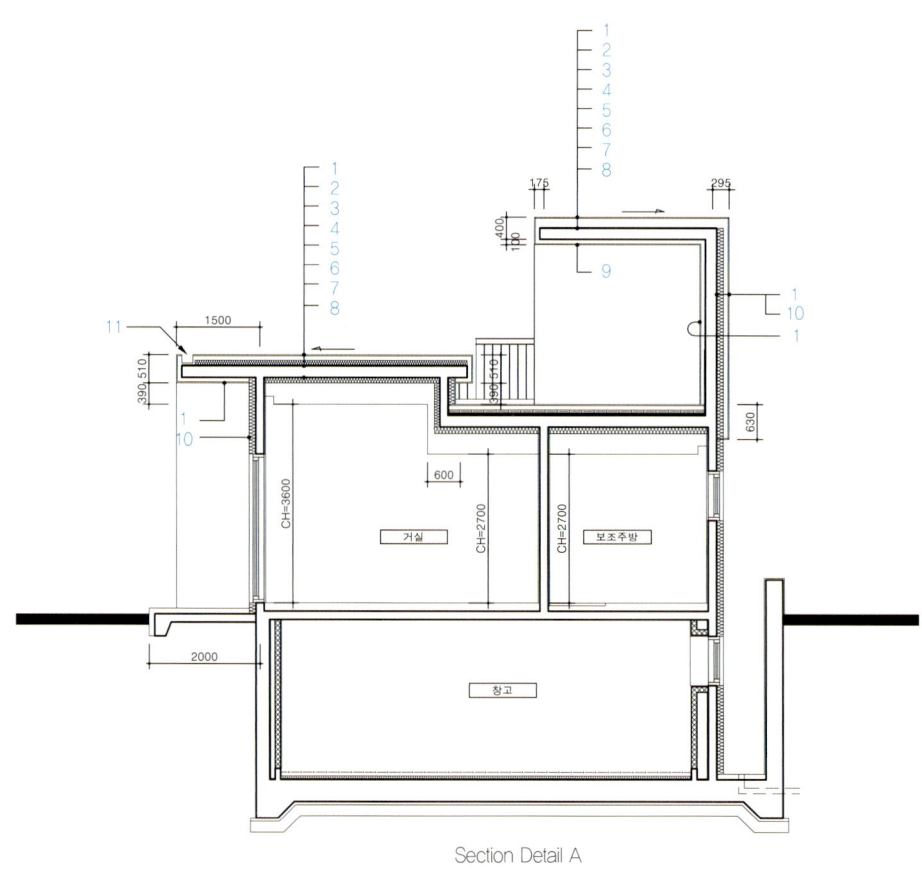

1. THK0.7 ZINC STANDING SEAM @430
2. GIWS WATERPROOF SHEET
3. 2T PLYWOOD
4. ST'L PIPE 40×40×1.6T
5. THK80 INSULATOR / T0.05 P.E FILM 2LAYERS
6. THK20 MORTAR
7. URETHAN FILM WATERPROOFING
8. CON'C TROWEL FINISH
9. T12 NATURAL WOOD FINISH
10. THK120 INSULATOR
11. W=200 GUTTER

1. THK0.7 징크 돌출이음 @430
2. GIWS 방수시트
3. 12T 내수합판
4. 각파이프 40×40×1.6T
5. THK80 단열재 / T0.05 PE필름 2겹
6. THK20 보호몰탈
7. 우레판 도막방수
8. 콘크리트쇠흙손마감
9. T12천연목재마감
10. THK120 단열재
11. W=200 거터

Section Detail A

1. THK0.7 ZINC STANDING SEAM @430	1. THK0.7 징크 돌출이음 @430
2. THK120 INSULATOR	2. THK120 단열재
3. T12 PLYWOOD	3. T12 천연목재마감
4. GIWS WATERPROOF SHEET	4. GIWS 방수시트
5. 12T PLYWOOD	5. 12T 내수합판
6. ST'L PIPE 40×40×1.6T	6. 각파이프 40×40×1.6T
7. THK80 INSULATOR / T0.05 P.E FILM 2LAYERS	7. THK80 단열재 / T0.05 PE필름 2겹
8. THK20 MORTAR	8. THK20 보호몰탈
9. URETHAN FILM WATERPROOFING	9. 우레판 도막방수
10. CON'C TROWEL FINISH	10. 콘크리트쇠흙손마감
11. EXPOSED MASS CONCRETE FINISH	11. 노출콘크리트 마감
12. THK120 EXTERIOR INSULATOR SYSTEM	12. THK120 외단열시스템

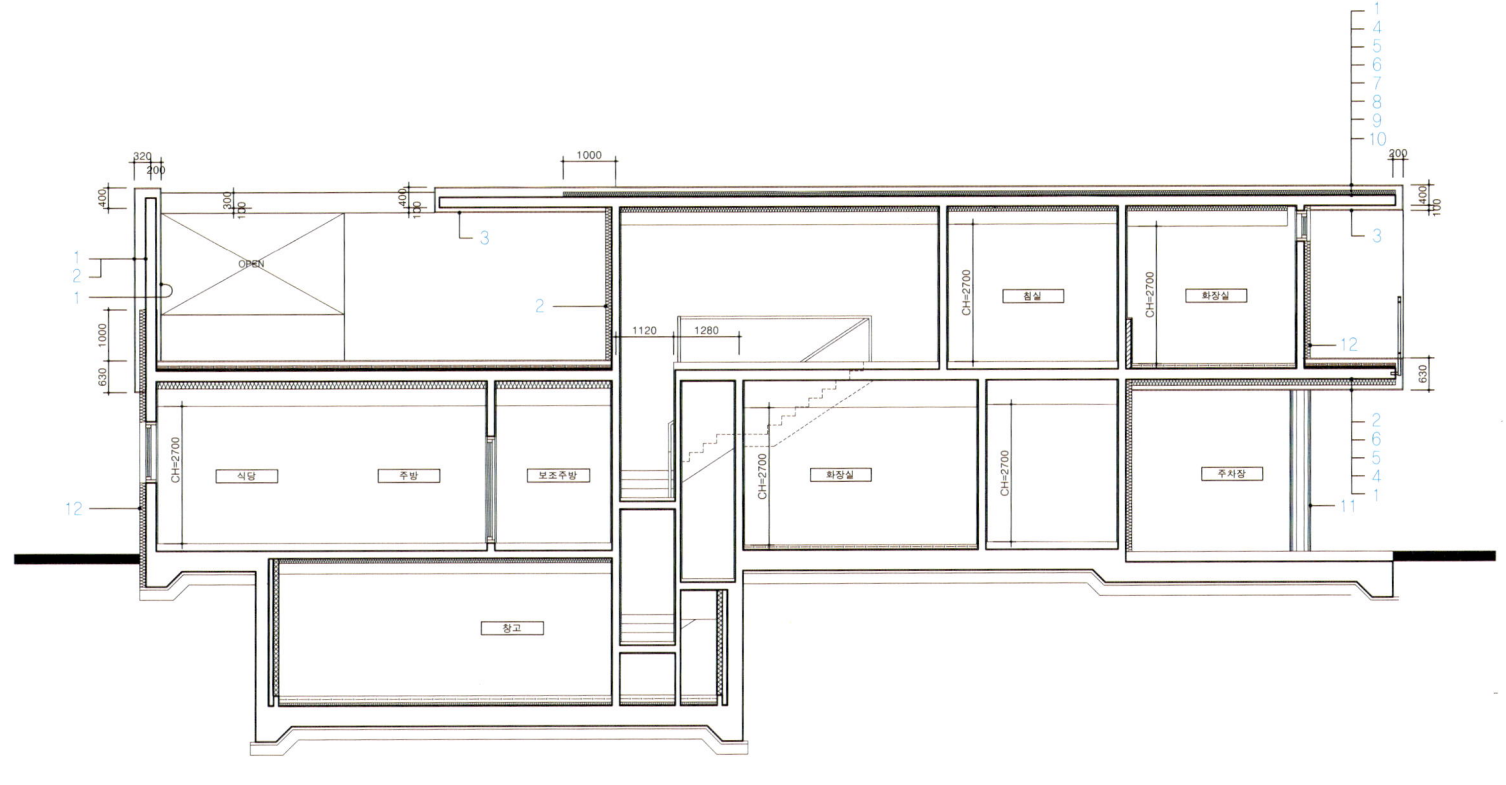

Section Detail B

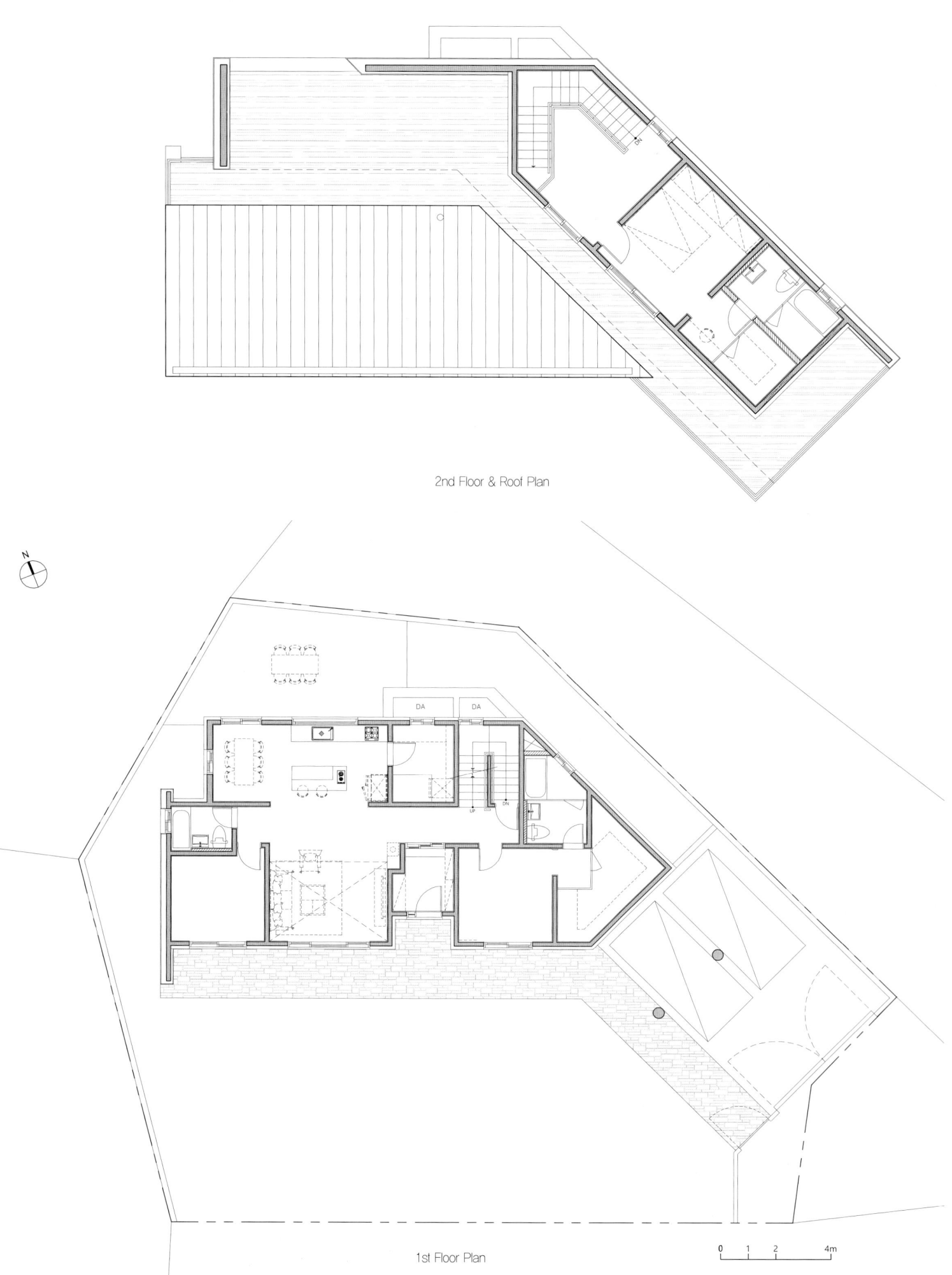

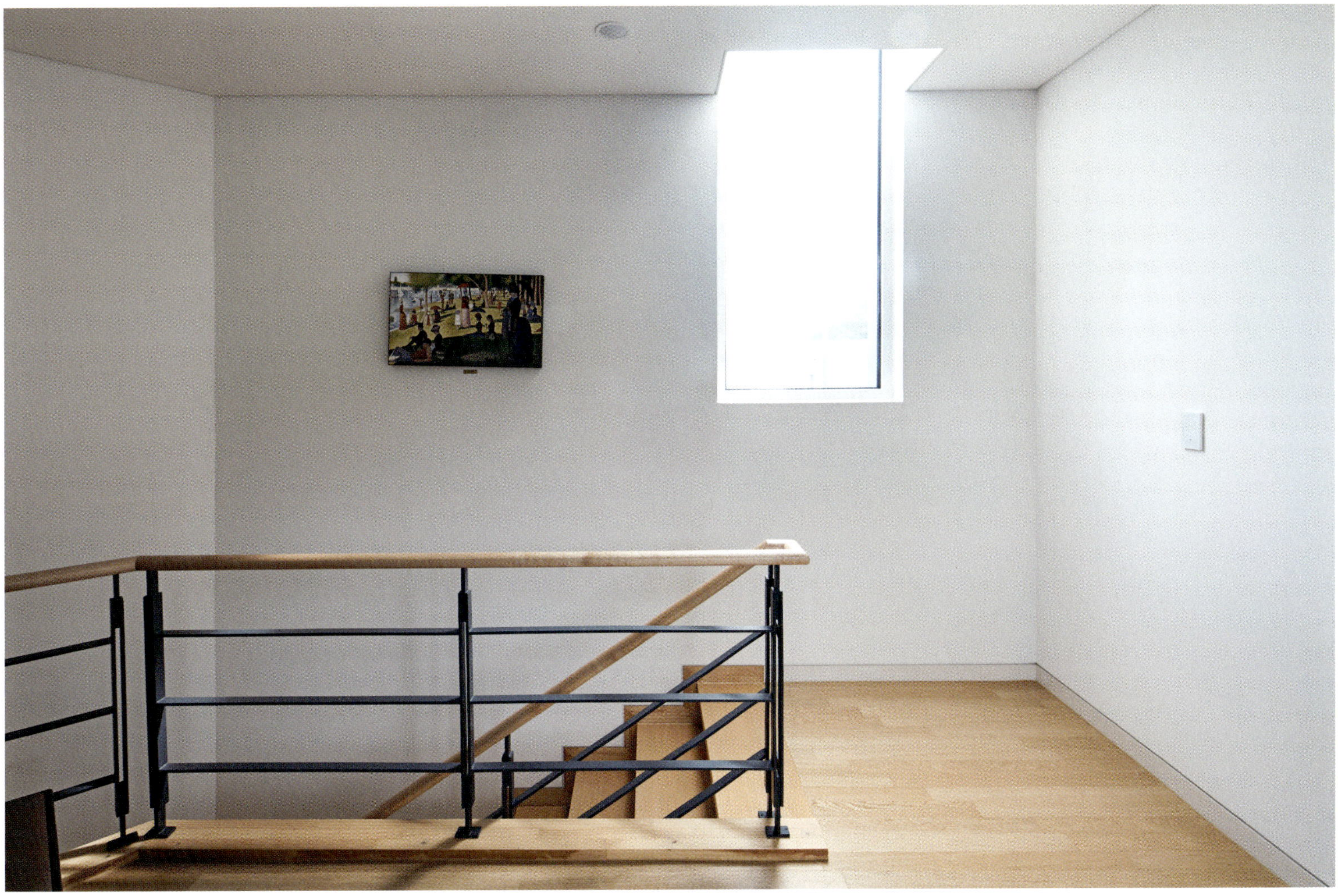

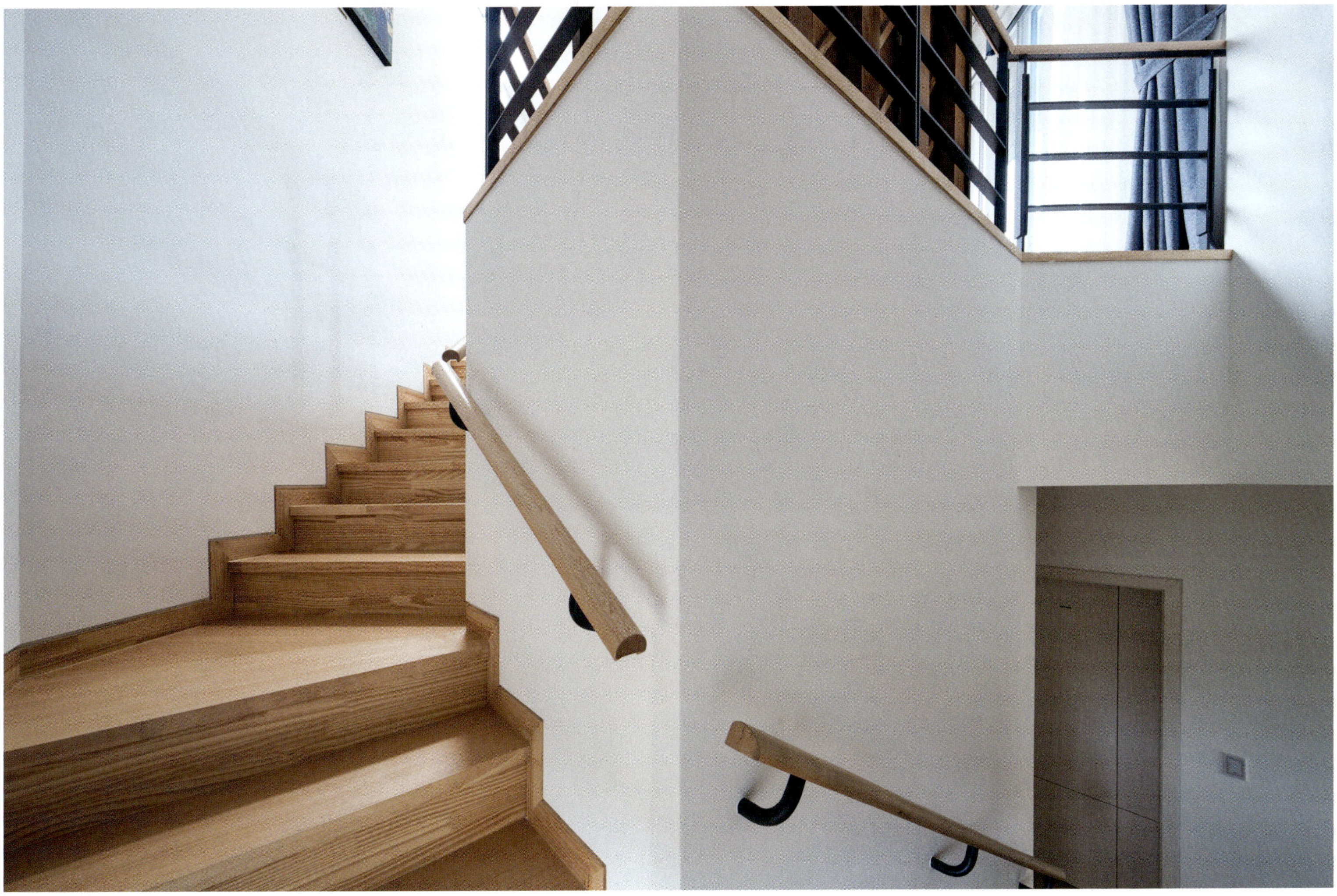

Green House Jinju, Korea

PLANO architects & associates

Architects PLANO architects & associates / Minsung Park, Wongil Lee, Geunhye Kim **Location** Jinju, Korea **Site Area** 301.7㎡ **Bldg. Area** 170.87㎡ **Gross Floor Area** 198.76㎡ **Bldg. Scale(Floor)** 2FL **Structure** Reinforced Concrete **Exterior Finish** Stucco **Photographer** Yongjun Choi

설계 플라노건축사사무소 / 박민성, 이원길, 김근혜 **위치** 경상남도 진주시 충무공동 45-10 **대지면적** 301.7㎡ **건축면적** 170.87㎡ **연면적** 198.76㎡ **규모** 지상2층 **구조** 철근콘크리트 **외부마감** 스타코 **사진** 최용준

Elevation

A Green house next to the park. The house has three gardens and three generations live together. The white house, like a drawing paper, has a park and a garden like a painting, and takes on a bluish color.

House in the park

The site faced the park to the north. It was a good location as a city-type house with limited green space. In order to maximize the advantages of the adjacent green space, we tried to connect it with the internal and external spaces of the site. The first floor created the flow of 'park-moss indoor garden-living room-front yard'. Centering on the common space connecting the north park and the south yard, private spaces and service spaces are arranged on both sides. The second floor also created a 'park-terrace-living room' flow to expand the interior and exterior spaces. Considering that the house will be built in the future, the south side has only a small ventilation window and has a skylight to secure light.

Three gardens

Rather than completing the architecture and adding landscaping, we considered architecture in which nature is the main character from the beginning. From the beginning of the design, three gardens with various spatial senses were planned through consultation with landscape experts.

1. Hackberry Garden: This is a lawn garden where you have to put on your shoes. It has a rough but natural taste that harmonizes with the external natural environment.

2. Grass Garden: It is a deck garden where you can go out barefoot. Although it is outside, it is connected to the interior and deck, so it can be used as a large space with the living room.

3. Moss Garden: This is the closest indoor garden that is always by your side. It is a garden of companion plants that you can communicate with at any time by stroking soft moss or fern with your bare hands.

1. ST'L FB-30×9 @200 / APP' PAINT
2. Ø12 ROUND BAR @200 / APP' PAINT
3. GRANITE CAPPING(FLASHING)
4. STUCCO ON MESH
5. URETHANE FILM WATERPROOFING
6. ANCHOR – PAINTING AFTER HANDRAIL WELDING
7. ANCHOR INSTALLATION
8. LAYING CRUSHED STONE
9. LANDSCAPING DRAINS
10. MOISTURE-PERMEABLE WATERPROOFING RESIN FOR ROOF
11. T80+T100 EXTRUSION METHOD INSULATION(XPS)
    NO. 1 OVERLAYING
12. REINFORCED CONCRETE STRUCTURE
13. DRIP HOME
14. WELDING AFTER ANCHOR INSTALLATION

1. ST'L FB-30×9 @200 / 지정도장
2. Ø12 환봉(원형봉) @200 / 지정도장
3. 화강석 두겁(후레싱)
4. 메쉬 위 스터코
5. 우레탄 도막방수
6. 앙카 – 난간 용접 후 도장
7. 앙카 설치
8. 쇄석 깔기
9. 조경용 배수판
10. 지붕용 투습방수지
11. T80+T100 압출법 단열재(XPS) 1호 겹쳐서 깔기
12. 철근콘크리트 구조체
13. 물끊기홈
14. 앙카 설치 후 용접

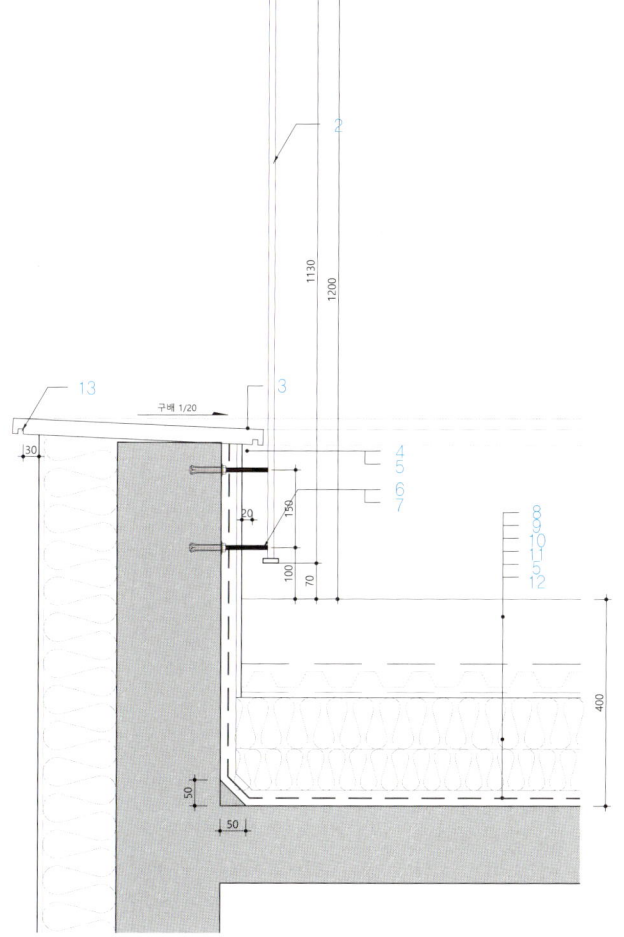

Handrail Section Dtail

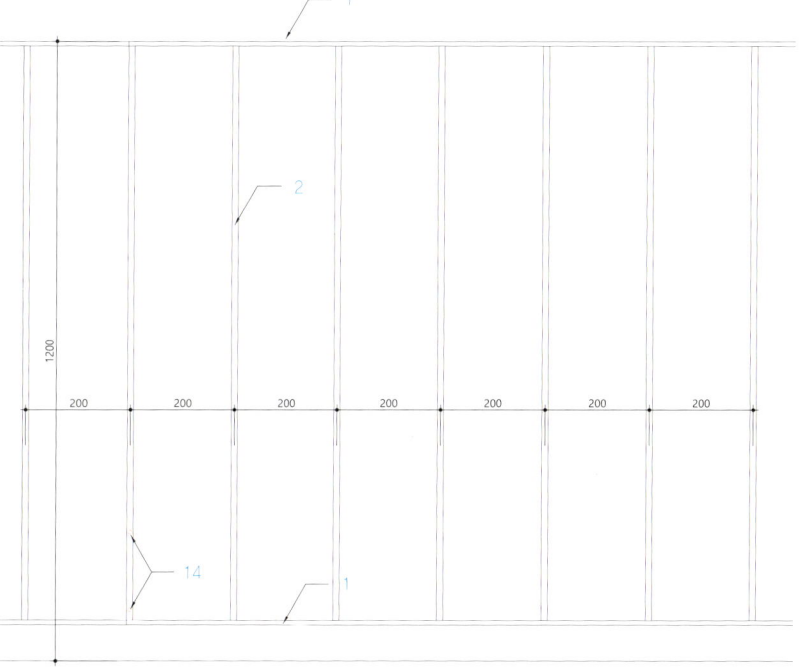

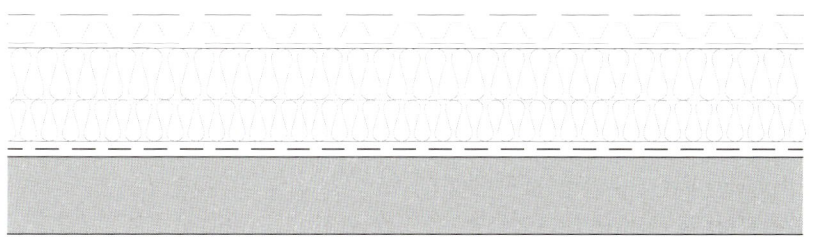

Handrail Elevation Detail

푸른집

공원 옆에 자리한 푸른집. 그 집에는 3개의 정원과 3대가 함께 산다. 도화지 같이 하얀 집은 공원과 정원을 그림처럼 담으며 푸른빛을 띤다.

공원에 닿은 집

대지는 북쪽으로 공원과 맞닿아 있었다. 녹지공간이 협소할 수밖에 없는 도심형 주택으로서 좋은 입지였다. 인접한 녹지의 장점을 극대화 시키고자 대지안의 내외부공간과 연결시키고자 노력했다. 1층은 '공원–이끼실내정원–거실–앞마당'의 흐름으로 만들었다. 북쪽 공원과 남쪽 마당 사이를 연결하는 공용공간을 중심으로, 양 옆으로는 개인공간과 서비스 공간을 배치하였다. 2층도 '공원–테라스–거실'의 흐름을 만들어 내외부공간을 확장시켰다. 남쪽은 추후 집이 들어설 것을 고려해 작은 환기창만 두고 천창으로 채광을 확보했다.

세개의 정원

건축을 완성하고 조경을 곁들인 것이 아니라 처음부터 자연이 주인공인 건축을 고민했다. 설계초기부터 조경전문가와 협의를 통해 공간적 거리감을 다양한 세 개의 정원을 계획했다.

1. 팽나무정원 : 신발을 신고 나가야 하는 잔디정원이다. 외부 자연환경과 어우러지는 거칠지만 자연스러운 맛이 있다.

2. 그라스정원 : 맨발로도 나갈 수 있는 데크정원이다. 외부지만 실내와 데크로 연결되어 있어 거실과 하나의 대공간으로도 이용한다.

3. 이끼정원 : 늘 곁에 있는 가장 가까운 실내정원이다. 보드라운 이끼나 고사리를 맨손으로 쓰다듬으며 언제든지 교감할 수 있는 반려식물들의 정원이다.

1. FILLING OF FLEXIBLE URETHANE FOAM
2. T20 GRANITE
3. T30 ADHESIVE MORTAR
4. MOSS PLANTING
5. VEGETATION SOIL
6. LANDSCAPING DRAINAGE BOARD
7. SHEET WATERPROOF
8. T5 MEMBRANE DRAIN PLATE
9. T100 EXTRUSION METHOD INSULATION(XPS) NO. 1
10. T0.1 PE FILM 2PLY
11. RUBBLE
12. T50 LEVELING CONCRETE

1. 연질우레탄폼 충진
2. T20 화강석
3. T30 붙임몰탈
4. 이끼식재
5. 식생토
6. 조경용 배수판
7. 시트방수
8. T5 멤브레인 배수판
9. T100 압출법 단열재(XPS) 1호
10. T0.1 PE필름 2겹
11. 쇄석
12. T50 버림콘크리트

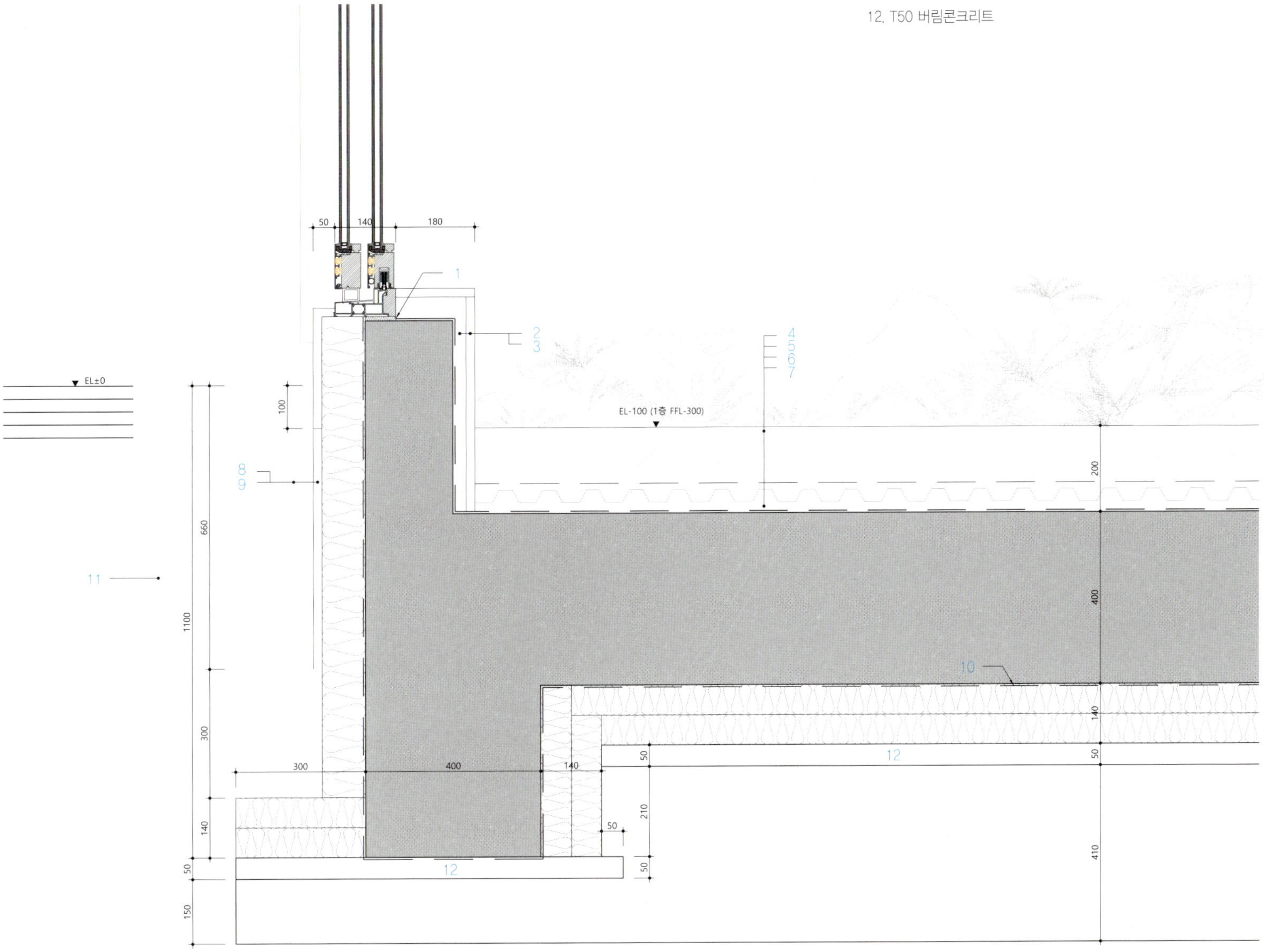

Foundation Section Detail

Section A

Section B

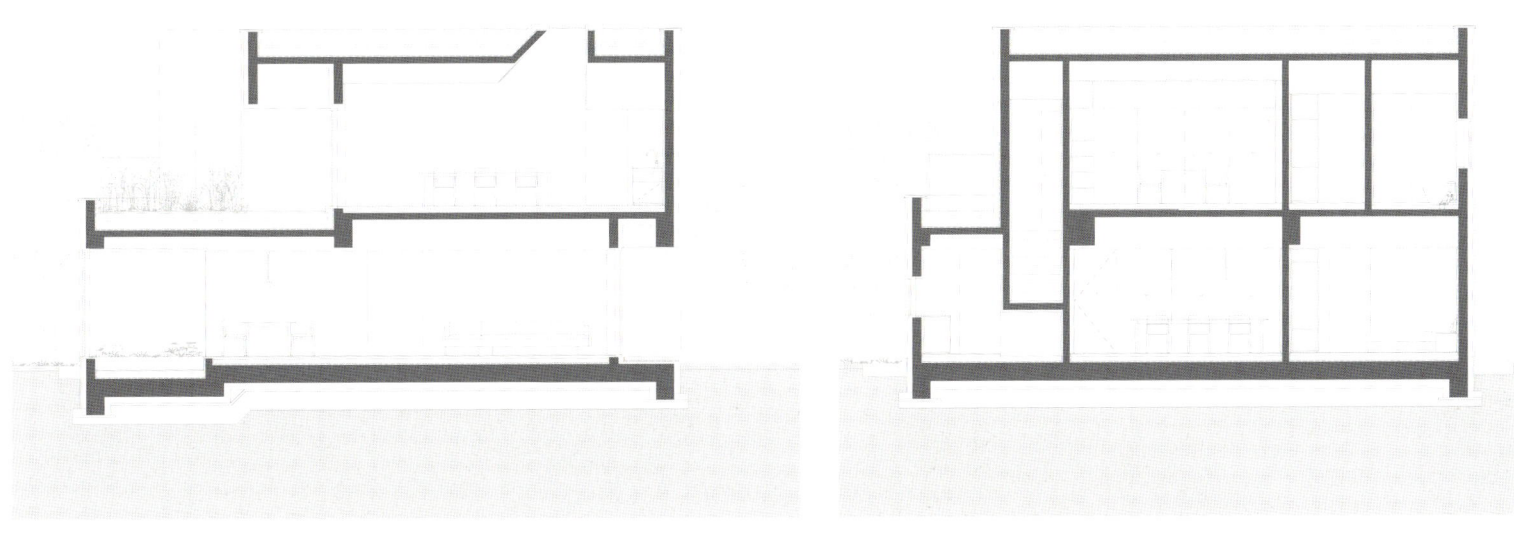

1st Floor Plan

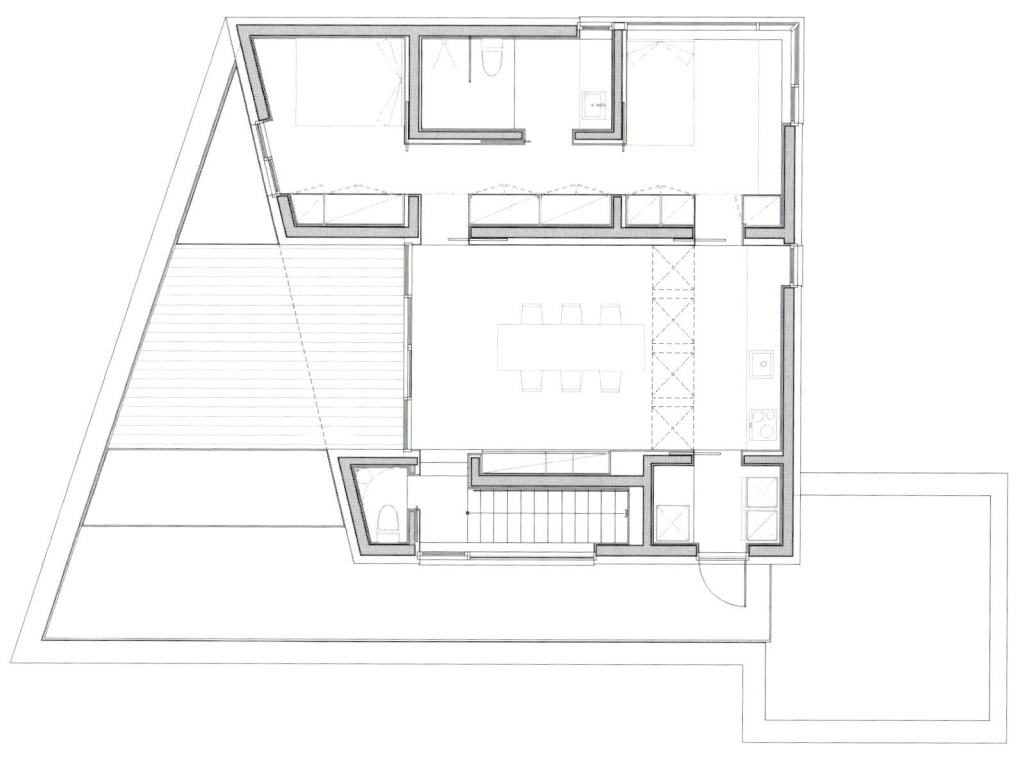

2nd Floor Plan

OYEONJAE Muju, Korea

Zerolimits Architects

Architects Zerolimits Architects / Jongseo Kim **Project Team** Yeonjin Lee, Eunseo Jo, Hyebin Lee, Yunha Um **Location** Muju, Korea **Site Area** 999.57㎡ **Bldg. Area** 173.68㎡ **Gross Floor Area** 190.75㎡ **Bldg. Scale(Floor)** 2FL **Structure** Reinforced Concrete + Plain Concrete **Max. Height** 7.2m **Exterior Finish** STO External Insulation System, Ceramic Wide Long Tile Black **Photographer** Jongseo Kim, Jueun Gong

설계 ㈜제로리미츠 건축사사무소 / 김종서 **설계참여자** 이연진, 조은서, 이혜빈, 엄윤하 **위치** 전라북도 무주군 설천면 두길리 1433 **대지면적** 999.57㎡ **건축면적** 173.68㎡ **연면적** 190.75㎡ **규모** 지상2층 **구조** 철근콘크리트 + 무근콘크리트 **최고높이** 7.2m **외부마감** 외단열시스템, 세라믹 와이드롱타일 블랙 **사진** 김종서, 공주은

Site Plan

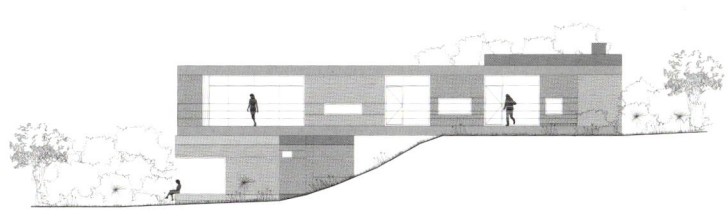

East Elevation

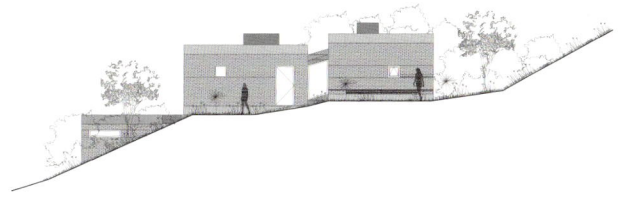

North Elevation

West Elevation

South Elevation

©Jongseo Kim

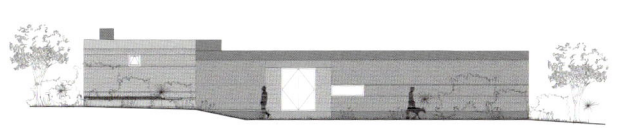

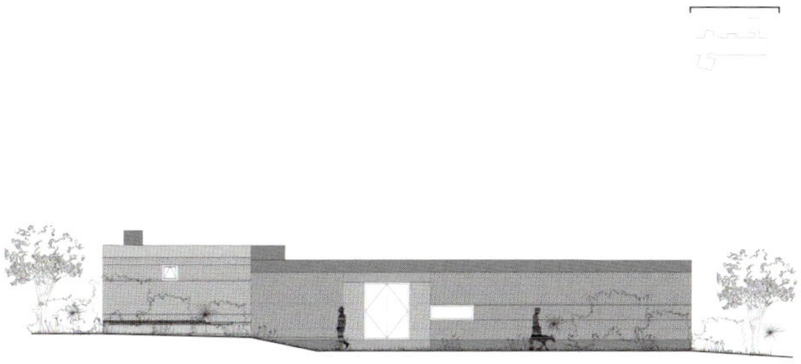

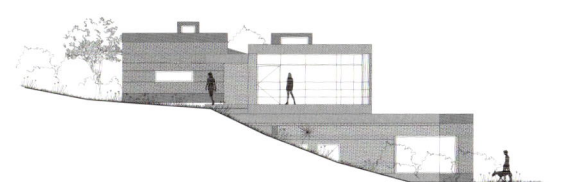

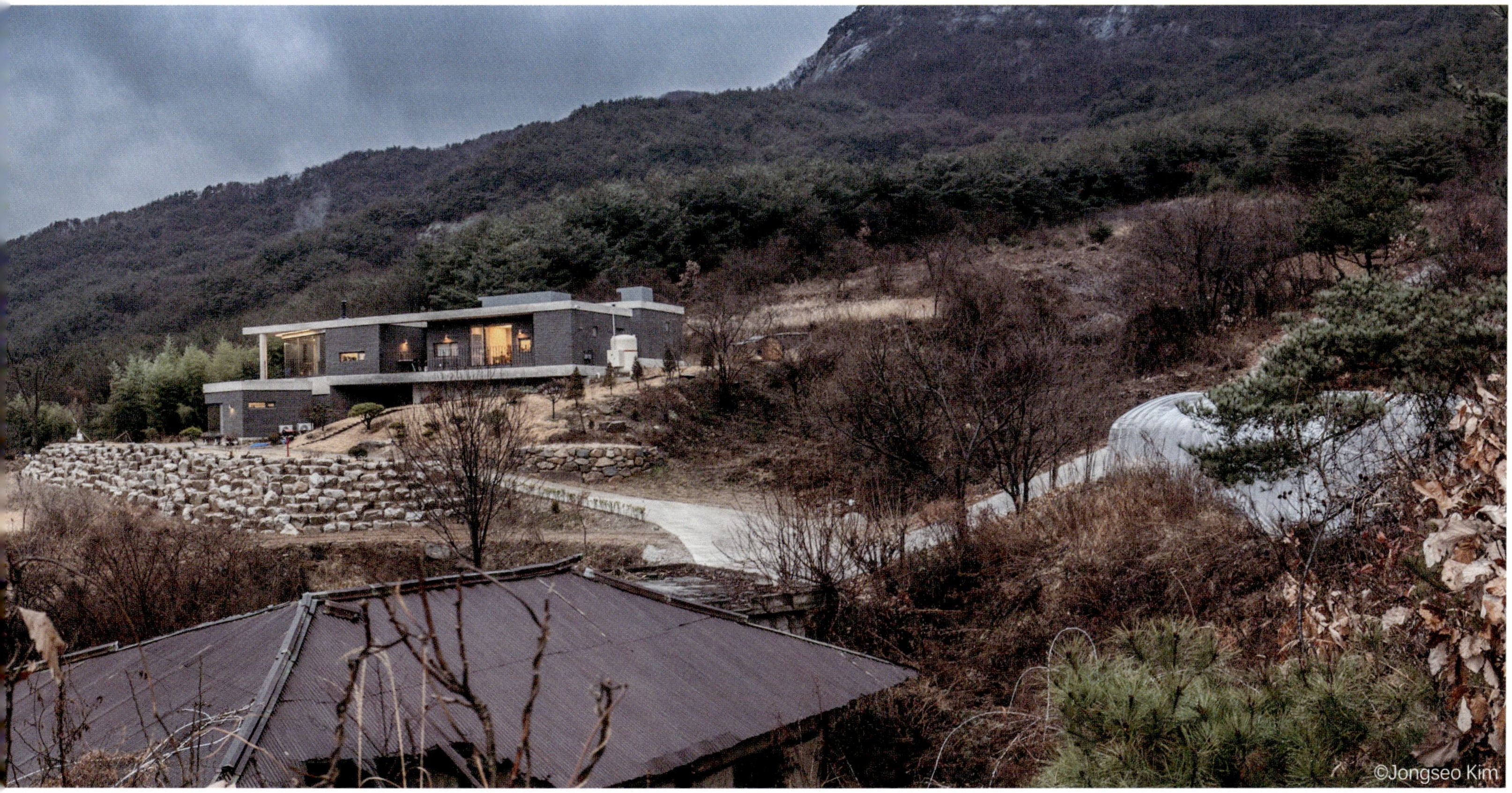

Earth

The owner of the building, whom I had been talking to for a long time, contacted me one day about two years ago. In the meantime, he purchased the land for his return to his hometown of Muju and was living in a nearby village with his family, and he also told me that he was worried about whether construction would be possible because the land he purchased was too steep. The first site I visited after a brief meeting in Seoul and half anticipation was a field located in the middle of a mountain with a fairly incline at a glance. It was a land with a cool view that can only be enjoyed on a sloped site where you can see the mountains and the village of Deogyusan at a glance. It is a slope that flows to the east, and the difference in height inside the site is about 6m. It had been used as a field for a long time, but now it has been neglected for a long time. Building is not impossible, so I told you not to worry, and on the way up to Seoul, the cool view of Deogyusan and old trees lingered in my head.

Arrangement

The basic principle of arranging the buildings in the basic condition of the site, which is the view from the south and northeast of the slope flowing to the east, was to ensure that the advantages of the site (view, smell) could be enjoyed as much as possible, and it was placed so that it could blend naturally without disturbing the surrounding topography. First of all, we considered the direction in which we can enjoy the mountains of Deogyusan to the maximum, which will show a different look from the old trees that have been guarded for a long time at the boundary of the land in the living room. There are yards at different levels for natural access to private spaces, adapting to the topography as much as possible, and these yards are arranged to connect with each interior space. The main mass of the second floor maintains the axis of the east direction, and the mass of the first floor is arranged along the south side of the Deogyusan view.

Floor plan

He asked for a space for the owner's family to live and a separate space to operate as a guest house. And although the two spaces are separated as separate spaces, he wanted them to be internally connected. The proposed space consists of three masses on the second floor, and each floor has an individual entrance using the characteristics of the slope. The second floor is the main residential space, where the owner couple and her two sons live. The living room, children's room, and dining kitchen, which are open to the south and east sides, are arranged horizontally, and windows are arranged according to the view of each space. The master room and the dressing room are composed of separate masses and are connected by a hallway. The first floor is a guesthouse for guests and consists of a separate living room, bedroom, and backyard. The first and second floors lead to the living room and are partitioned by doors that can be opened as needed.

Space configuration

Contemplation and response to the condition of the site, which is a slope with a difference of more than 3m, was an important part of determining many aspects of this plan. It was not easy to find a composition that could make the most of the architectural possibilities that are only possible on sloping land, not flat land, secure the space requested by the building owners, and adapt as much as possible without harming the topography. There also had to be a plan that would allow efficient construction.

First of all, the axis of the main building is arranged as much as possible in the same direction as the direction of the inclined flow so that it can adapt to the flow of the existing terrain. made to follow For this reason, the mass was superimposed on two different axes to secure the scent and view according to the characteristics of each space.

1. THK180 단열재(가등급)	1. THK180 INSULATION('A' GRADE)
2. THK9.5 석고보드2겹	2. THK9.5 GYPSUM BOARD 2PLY
3. 지정색 도장	3. APP' COLOR PAINTING
4. THK8 지정 온돌마루	4. THK8 APP' ONDOL FLOOR
5. THK50 몰탈(Ø15 PIPE 삽입)	5. THK50 MORTAR(INSERT Ø15 PIPE)
6. THK60 기포콘크리트	6. THK60 AERATED CONCRETE

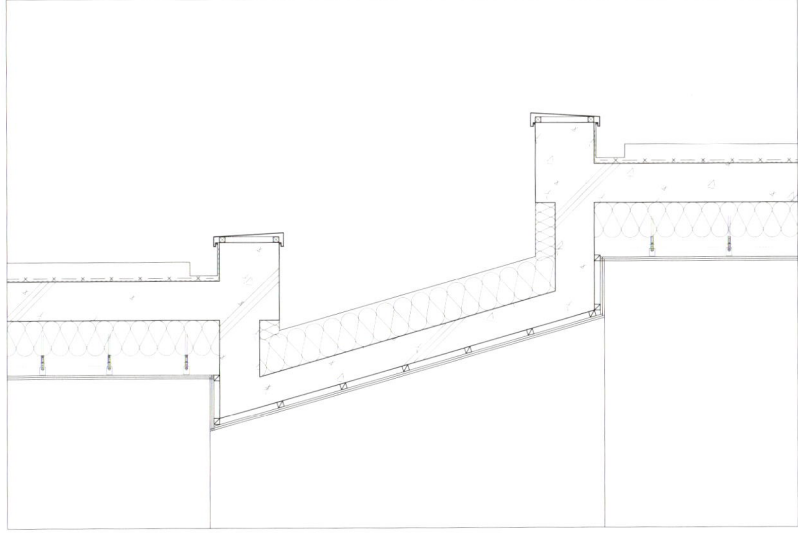

Detail A

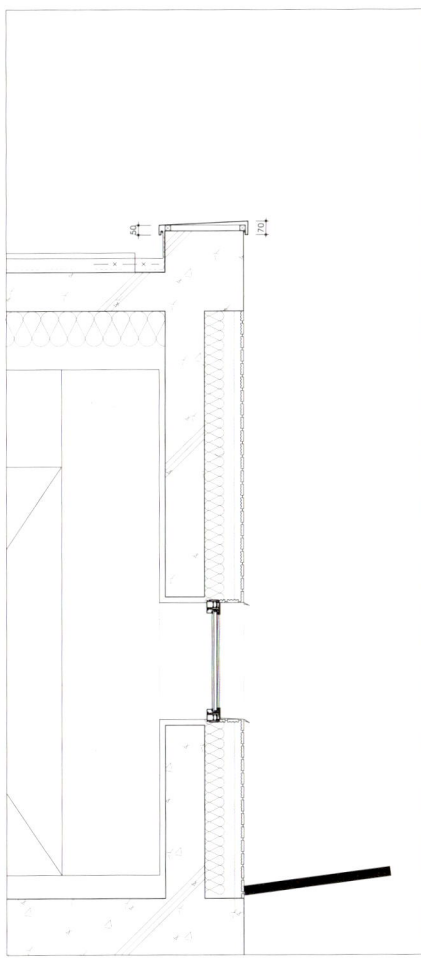

Detail B

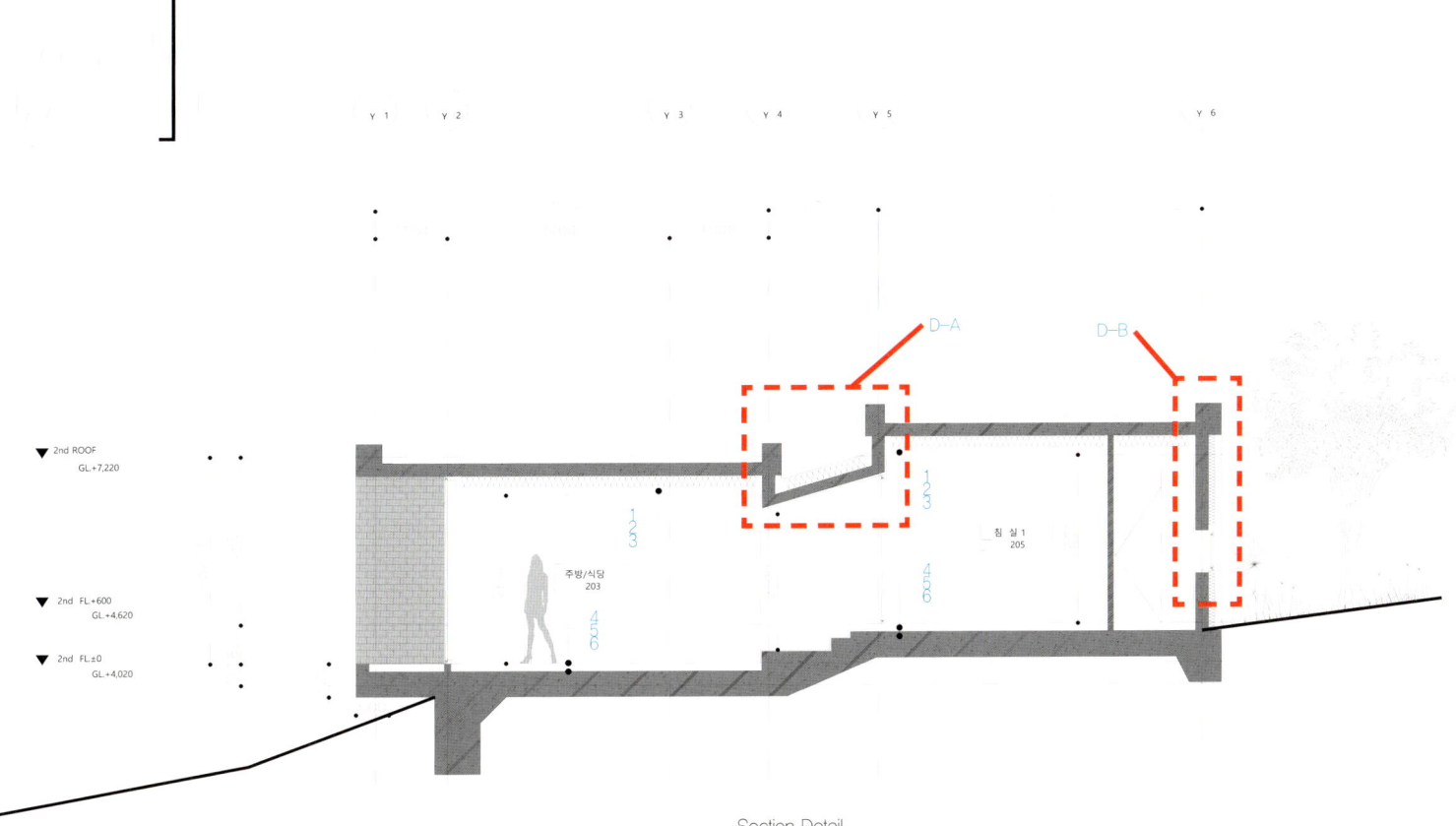

Section Detail

오연재

대지

오래 전 상담통화를 하셨던 건축주 분께서 2년쯤 지난 어느날 연락해 오셨다. 그동안 고향인 무주로 귀촌 하실 땅을 구입하시고 가족들과 인근마을에서 거주하고 계신 상황이었고, 구입한 대지가 너무 급한 경사지라 건축이 가능할지 걱정이 된다는 말씀도 전해주셨다. 서울에서 간단한 미팅을 하고 걱정반 기대반으로 처음 방문한 대지는 한눈에 보기에도 꽤 경사가 있는 산골의 중턱에 위치한 밭이었다. 덕유산의 산세와 마을의 전경이 한눈에 들어오는 경사지의 대지에서만 누릴 수 있는 시원한 전망을 가지고 있는 땅이었다. 동쪽으로 흐르는 경사면으로 대지 내부의 높이 차이가 6m정도 되었다. 밭으로 오랫동안 쓰였으나 현재는 방치되던 오래된 상태였고 눈에 뛰는 것은 오래동안 여기를 지켰을 고목 한그루가 대지의 경계를 지키고 있었다. 건축이 불가능 하지 않으니 걱정마시라는 말씀을 드리고 서울로 올라오는 길에 시원한 덕유산의 전경과 오래된 고목이 머릿속을 맴돌았다.

배치

동쪽으로 흐르는 경사 남쪽과 동북쪽의 전망이라는 대지의 기본 조건에서 건물 배치의 기본 원칙은 최대한 대지가 가지고 있는 장점(조망, 향을 누릴 수 있고 주변 지형을 거스르지 않고 자연스럽게 어우러질 수 있게 배치하려 했다. 우선 거실은 대지경계에서 오래 지키고 있는 고목과 사철 다른 모습을 보여줄 덕유산의 산세를 최대한 누릴 수 있는 방향을 고려하였다. 지형에 최대한 순응하며 자연러운 사적공간으로의 접근을 위한 각기 다른 레벨의 마당이 있으며 이 마당은 각 내부 공간들과 연결되도록 배치되었다. 2층 주매스는 정동향의 축을 유지하고 있으며 1층 매스는 덕유산 전망의 남향을 따라 배치되어있다.

평면

건축주 가족이 거주할 공간과 게스트 하우스로 운영하실 별도의 공간을 요구하셨다. 그리고 2개의 공간은 별도의 공간으로 분리되어 있되 내부적으로 연결되길 원하셨다. 제안된 공간은 2층에 3개의 매스로 구성되어 있으며 경사지의 특성을 이용하여 각층의 개별 출입구를 가지고 있다. 2층이 메인 주거공간으로 건축주 부부와 아들 2명이 거주하는 곳이다. 남쪽과 동쪽 2면으로 개방된 거실과 아이들방, 식당 주방이 횡으로 배열되어 있으며 각 공간의 전망에 따라 창들이 배치되어있다. 마스터룸과 드레스룸은 별도의 매스로 구성되어 있으며 복도로 연결되어있다. 1층은 손님들을 위한 게스트하우스로 별도의 거실, 방, 뒷마당 등이 구성되어 있다. 1층과 2층은 거실로 이어져 있으며 필요에 따라 개방되는 문으로 구획되어있다.

공간구성

3m이상의 차이가 나는 경사지라는 대지의 조건에 어떤 태도를 취할 것인가에 대한 고민과 대응이 이번계획의 많은 사항들을 결정짓는 중요한 부분이었다. 평지가 아닌 경사지의 땅에서만 가능한 건축적 가능성을 최대한 활용하고 건축주 분들이 요구하신 공간을 확보하며 최대한 지형을 해치지 않고 순응할 수 있는 구성을 찾는 것은 쉬운 일이 아니었다. 거기에 효율적인 시공이 가능한 계획이기도 해야 했다.

우선 주동의 축을 경사 흐름의 방향과 최대한 동일하게 배치하여 기존지형 흐름에 순응 하게 하였고 기능상 효율적인 접근성이 필요한 게스트 하우스 매스를 진입동선에서 가장 가까운 곳에 배치하고 또다른 요소인 조망과 향에 따르는 축에 따르도록 하였다. 이런 이유로 2개의 서로 다른 축으로 매스가 중첩되어 각 공간의 성격에 따른 향과 조망을 확보하였다.

Section A

Section B

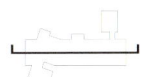

Section C

Section D

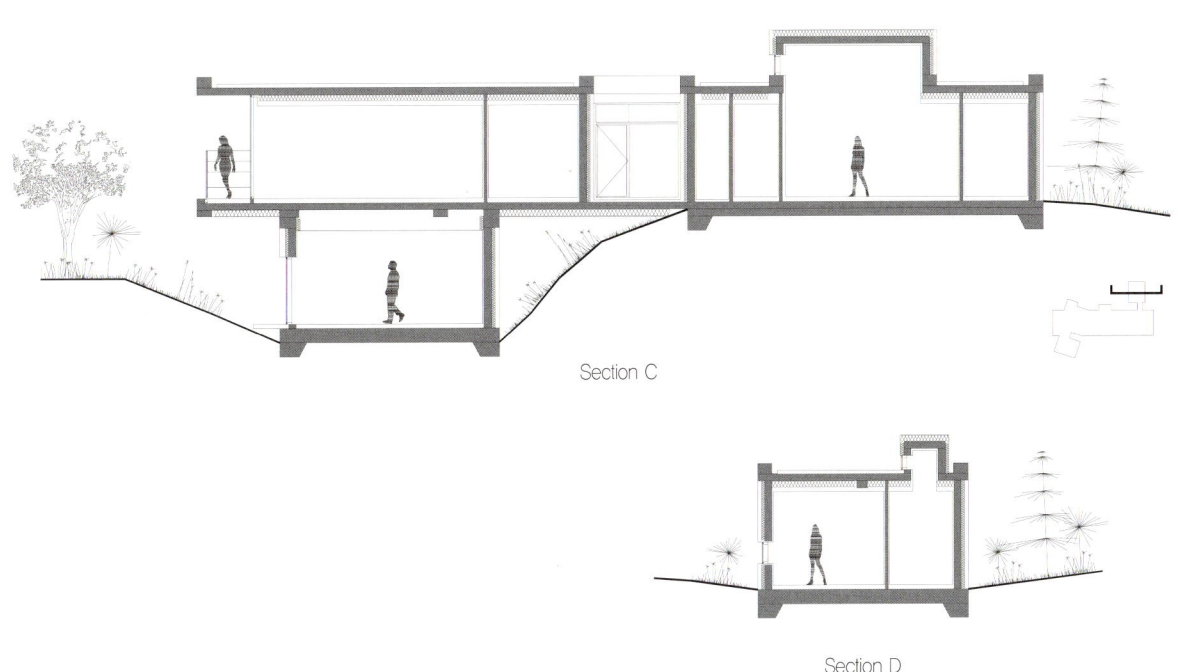

1. ENTRANCE
2. LIVING ROOM & KITCHEN
3. BED ROOM
4. BATHROOM
5. LIVING ROOM
6. LIVING ROOM 2
7. BED ROOM 2
8. ENTRANCE 2
9. DECK
10. BATHROOM 2
11. KITCHEN & DINING
12. UTILITY ROOM
13. BED ROOM 3
14. BATHROOM 3
15. DRESS ROOM
16. DECK 2

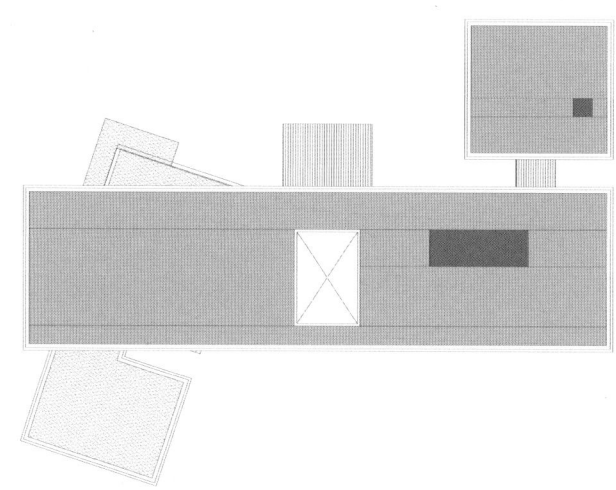

Roof Plan

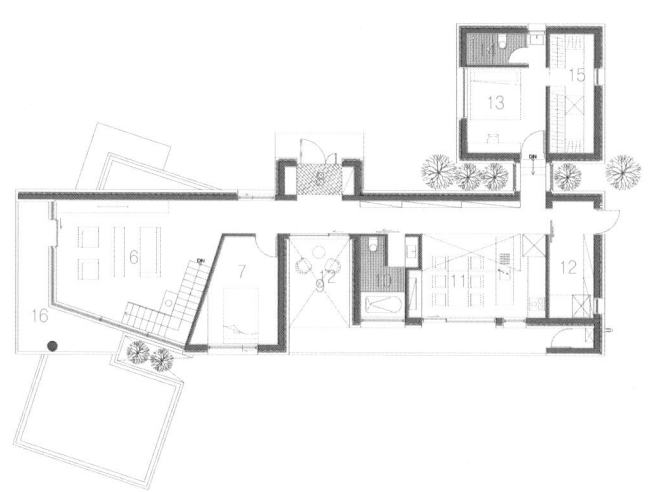

2nd Floor Plan

1st Floor Plan

©Jueun Gong

©Jueun Gong

213

Sangseon-won Namyangju, Korea

We Architects

Architects We Architects / Minchoul Sin **Location** Namyangju, Korea **Site Area** 876㎡ **Bldg. Area** 251.3㎡ **Gross Floor Area** 440.24㎡ **Structure** Reinforced Concrete **Exterior Finish** Roof_Zinc Panel, Wall_Exposed Concrete, Corrugated Brick, Q–Block, Deck_LG Synthetic Wood **Photos Offer** We Architects

설계 위종합건축사사무소 / 신민철 **위치** 경기도 남양주시 화도읍 **대지면적** 876㎡ **건축면적** 251.3㎡ **연면적** 440.24㎡ **구조** 철근콘크리트 **외부마감** 지붕_징크판넬, 벽_노출콘크리트, 파벽돌, 큐블럭, 데크_LG 합성목재 **사진제공** 위종합건축사사무소

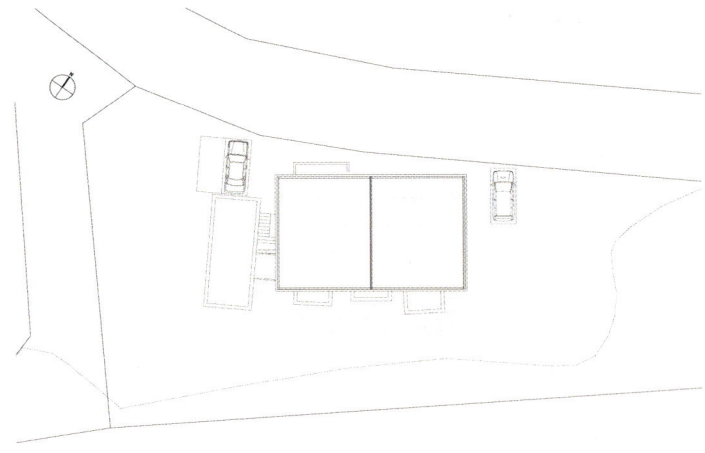

Site Plan

Front Elevation

Left Elevation

Rear Elevation

Right Elevation

When receiving a design request for a house, after hearing and discussing the owner's thoughts, the house is named and design work begins. Sangseonwon means 'the best good is like water' in the words 'sangseon medicinal water' in the Lao-tzu Tao sutra. The name of the site contains the meaning of looking at the Bukhangang(River), the core landscape of the site, and using the 'Water Way' as an indicator of life. The yard you see when you open the gate is divided into three main sections: the entrance yard on the same level as the first floor, the basement-level front yard with a small pool in front, and the back yard with the main sauna. When you reach the entry yard, you will be greeted by a handsome old pine tree that was originally located with the wide Bukhangang(River)in the background. It seems to silently explain to the visitor that the building is not the main character, but the wide open Paldangho Lake.

Space for guests and landlords

It has a topography that gradually lowers from the long access road at the back to the Bukhangang(River)in the southeast direction, and the site spread out toward the river naturally creates an architecture that conforms to the topography. The entrance from the road is on the first floor, where the entrance and living room are located, and in the courtyard on the basement floor facing the Bukhangang(River) there is a small pool, a public sauna and a banquet hall connected to it, so that guests can be welcomed from the outside space even in the courtyard. The guest room on the 2nd floor has a balcony in front, which acts as a buffer with the outside space and is used as a stage to enjoy the river breeze. The entrance is located on the right side of the yard where you can see from the large kitchen window facing the pine tree. The owner of the house who likes to cook said he wanted to cook food while watching his acquaintances come and wanted the front door to be visible from the kitchen. Inside, the dining room and living room on the first floor, which are the spaces for the guest and the owner, occupy a large space. Each room draws the river water into the house through a large window at the end of a long corridor where you can see the Bukhangang(River)like a telescope. The stairs facing Sunken Garden have long windows to show the movement of upper and lower space from the basement to the second floor. The stairs are in an open form and are designed to enjoy the surrounding scenery by going up and down the stairs rather than moving up and down, which is the general purpose of the stairs.

See-through and white porcelain moon jar

While designing and constructing this house, I tried to embody the dignity of hanok in the present age. In the selection of architectural materials, the exterior is finished with corrugated brick, a material that has a natural texture and color rather than a standardized and uniform size, and the scenery of the Bukhangang(River)in the front is filtered through a sari cloth to draw in the interior as a car scene. Blocks were constructed in a manual way to create irregular spaces between blocks, giving them a natural, handmade sensibility. Such a niche space resembles the crack of a moon jar among my favorite white porcelain of Joseon porcelain, and it is naturally formed by the cracking of the glaze by enduring the heat in the kiln over a thousand degrees. Sweat made it natural. The two-story open space, which would otherwise appear too large, was made as close to a human scale as possible by placing a visual passage with a large frame. The plane is open towards the river. As if to spread arms toward the river rather than a rectangular room at a right angle, the planes tried to expand the perspective by creating a subtle diagonal line toward the river. The effect created a greater amount of openness towards the river than the depth of the plane.

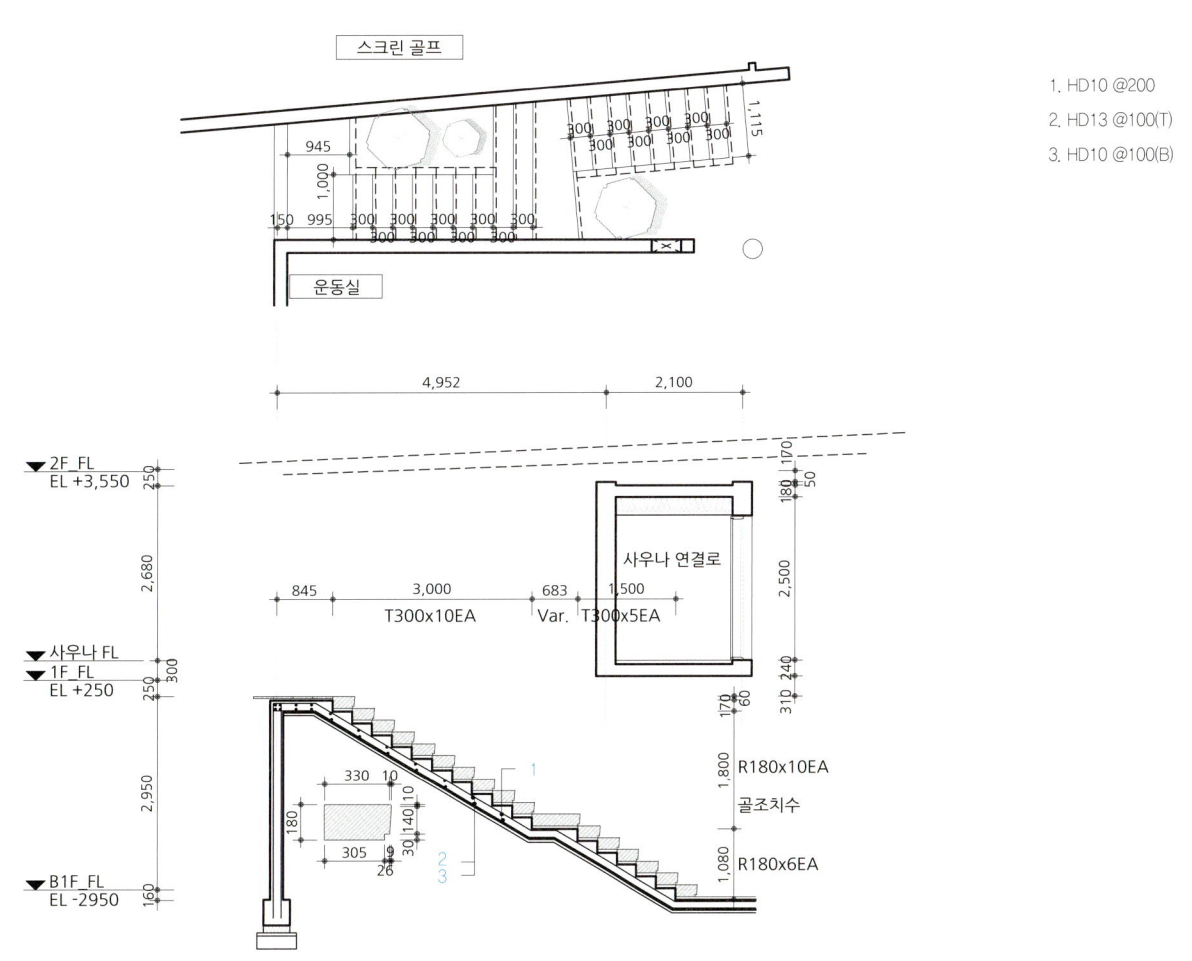

Outdoor Stairs Detail

1. HD10 @200
2. HD13 @100(T)
3. HD10 @100(B)

상선원(上善院)

주택 설계 의뢰를 받으면 건축주의 생각을 듣고 상의한 후, 집 이름을 짓고 디자인 작업이 시작된다. 상선원은 노자 도덕경에 나오는 '상선약수(上善若水)'라는 말로 '최상의 선은 물과 같다'는 뜻이다. 당호에는 사이트의 핵심 풍경인 북한강을 바라보며 '물의 도'를 인생 지표로 삼겠다는 의미가 담겨있다. 대문을 열고 들어가면 보이는 마당은 1층과 같은 높이의 진입마당, 전면에 조그만 풀장이 있는 지하층 높이의 앞마당, 안방 사우나가 있는 뒷마당으로 크게 세 부분으로 나누어진다. 진입마당에 이르면 본래부터 자리 잡고 있던 잘생긴 오래된 소나무가 넓은 북한강을 배경으로 반갑게 맞이한다. 이곳은 건물이 주인공이 아니고 넓게 펼쳐진 팔당호가 주인임을 방문자에게 소리 없이 설명하는 듯하다.

게스트와 집주인의 공간

뒤편에 길게 접한 진입로로부터 점차 남동 방향의 북한강으로 낮아지는 지세를 갖고, 넓게 강을 향해 펼쳐진 대지는 자연스럽게 지형에 순응하는 형태의 건축을 만들었다. 도로에서 진입하는 부분은 1층으로 현관과 거실이 위치하고, 북한강 쪽인 지하층 안마당에는 작은 풀장과 그것에 연계된 공용 사우나. 연회장을 두어 안마당에서도 손님들을 외부공간에서 맞이할 수 있게 했다. 2층의 게스트 룸은 전면에 발코니를 두어 외부 공간과 완충 역할을 하며 강바람을 감상하는 무대로 활용된다. 현관은 마당 우측에 소나무와 마주한 커다란 주방 창에서 보이는 곳에 자리한다. 요리를 좋아하는 집 주인은 지인들이 오는 것을 보면서 음식을 만들고 싶다며 부엌에서 보이는 위치에 현관이 있기를 원했다. 내부는 게스트와 집주인 공간인 1층 식당과 거실이 크게 자리를 잡는다. 각 실들은 북한강을 망원경처럼 볼 수 있는 긴 복도의 끝에 달린 큰 창을 통하여 강물을 집 안으로 끌어들인다. 선큰 가든과 접한 계단은 긴 창문을 두어 상하공간의 이동을 지하층에서부터 2층까지 보여준다. 계단은 오픈된 형태로 일반적 계단 목적인 상하 이동이 아닌 계단을 오르내리며 주변 경치를 감상하도록 구성됐는데, 이는 마르셀 뒤샹의 '계단을 내려오는 나부'에서 영감을 받았다.

시스루와 백자 달항아리

이 집을 설계 시공하면서 한옥의 고격이 갖는 품격을 현 시대에 구현하기 위한 노력을 해보았다. 건축적인 재료의 선정에 있어서도 규격화되고 획일적인 크기를 갖고 있는 것보다는 자연스러운 질감과 색감을 갖는 재료인 파벽돌로 외장을 마무리하고, 전면의 북한강 풍경을 사리 천에 걸러서 차경으로 내부에 끌어들이는 역할을 하는 블록을 수공업적인 방법으로 시공하여 블록 틈새 공간을 불규칙하게 만들어 자연스러운 느낌의 핸드메이드적인 감성을 주었다. 그런 틈새 공간은 내가 좋아하는 조선 도자기 백자 중에 달항아리의 균열과도 닮아 있으며, 그것은 천도가 넘는 가마 속 열기를 견디며 유약이 갈라지며 자연스럽게 형성된 것이고, 이 집의 틈들은 뜨거운 여름날 뙤약볕에서 일한 조적공들의 땀이 자연스러움을 만든 것이다. 자칫하면 너무 크게 보일 수 있는 두개 층이 오픈된 공간을 대형 프레임의 시각적 통로를 두어 최대한 휴먼 스케일에 가깝게 만들 수 있었다. 평면은 강을 향해 열려 있다. 직사각으로 된 방이 아닌 강 쪽으로 팔을 벌릴 것처럼 평면들은 강 쪽으로 미묘한 사선을 만들어 시각의 확장을 꾀했다. 그 효과는 평면의 깊이보다 더 많은 양의 개방감을 강 쪽으로 만들었다.

조선 시대의 뛰어난 유물로 반듯이 꼽히는 것이 달항아리이다. 그 이유는 절제와 담박함으로 빚어낸 순백의 빛깔과 둥근 조형미에 있다. 그러나 많은 전문가들도 추측만 할 뿐이지 정확한 쓰이는 용도를 알지 못한다. 건축가는 이와 같은 '집'이란 그릇을 만드는 사람일 뿐이고, 그 집은 사용하는 사람의 인생을 투영하며 삶을 담는 그릇이 되는 것이다.

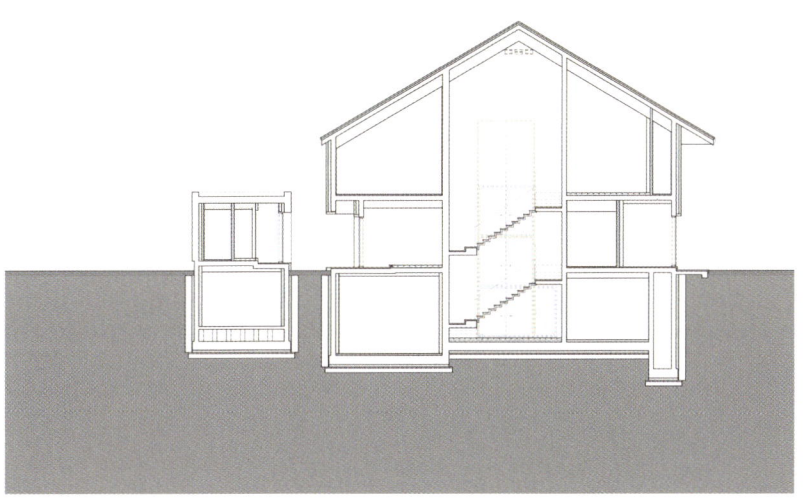

Section A

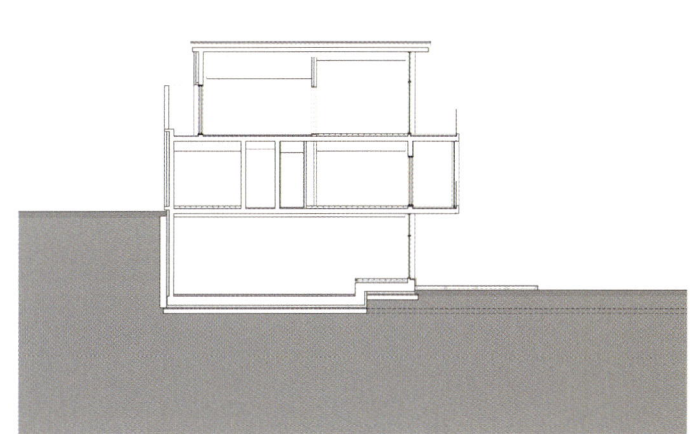

Section B

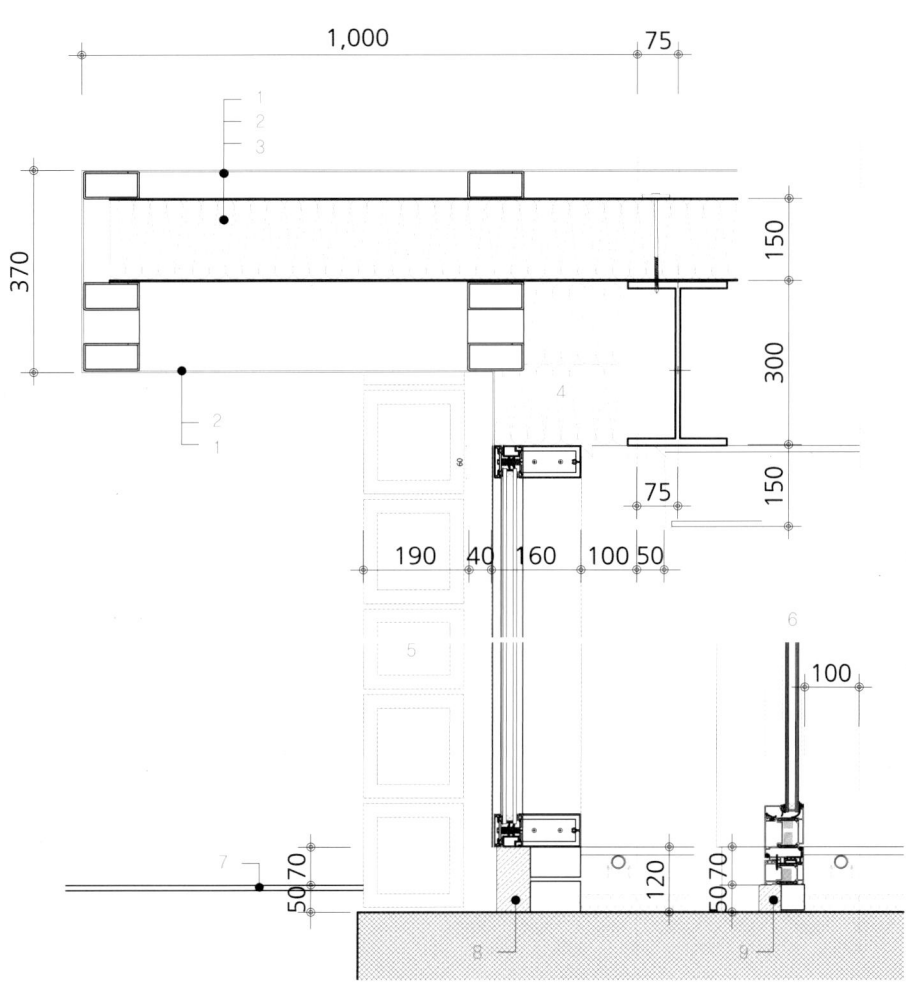

Front Eaves & Floors Detail

1. THK0.5 ZINC STEEL PLATE(@370)
2. ロ-100×50 SQUARE PIPE
3. THK150 FLAME RETARDANT PANEL
4. URETHANE FOAM FILLING
5. 190×190×190 Q-BLOCKS
6. B1F & 2F SLIDING DOOR
7. COMPOSITE WOOD DECK
8. ARTIFICIAL MARBLE 60×120
9. ARTIFICIAL MARBLE 40×50

1. THK0.5 징크강판(@370)
2. ロ-100×50 사각파이프
3. THK150 난연판넬
4. 우레탄폼 충진
5. 190×190×190 큐블럭
6. B1F & 2F 여닫이 문
7. 합성목재 데크
8. 인조대리석 60×120
9. 인조대리석 40×50

▼ B1F_FL

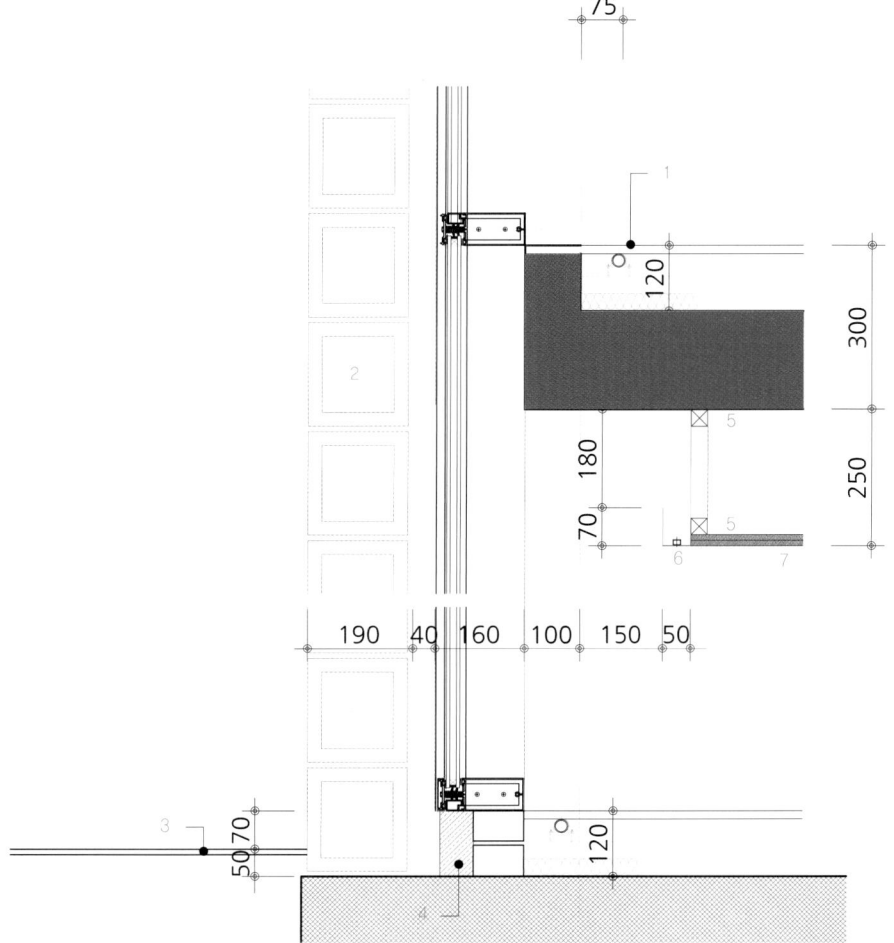

Indirect Light Box Detail

1. APP' FINISH ON THK120 PANEL HEATING
2. 190×190×190 Q-BLOCKS
3. COMPOSITE WOOD DECK
4. ARTIFICIAL MARBLE 60×120
5. 30×30 WOOD
6. 1T GALVANIZED STEEL PLATE
7. GYPSUM BOARD 2PLY

1. THK120 판넬히팅 위 지정 마감
2. 190×190×190 큐블럭
3. 합성목재 데크
4. 인조대리석 60×120
5. 30×30 목재
6. 1T 갈바철판
7. 석고보드 2겹

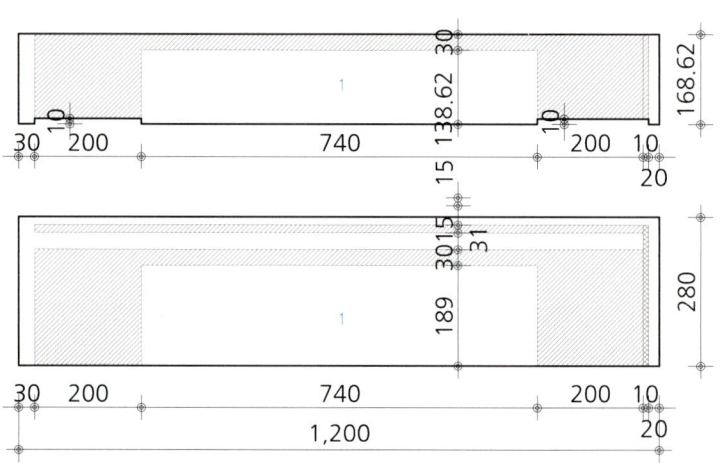

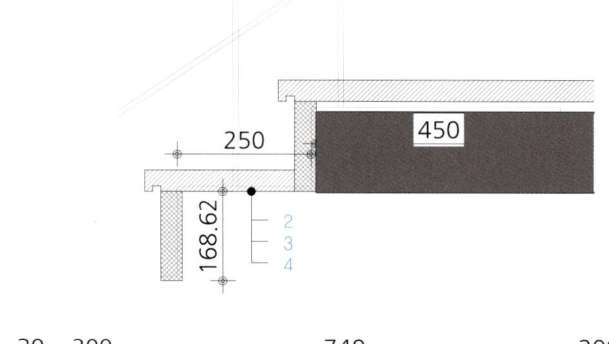

1. T40 WOODEN STAIRCASE - RISER
   DIG THE HATCHED PART T10
2. T40 WOODEN STAIRCASE
3. T1 RUBBER PLATE
4. ENAMEL PAINT ON T9 F.B

1. T40 목재 계단판-챌판
   빗금 부분 T10 파내기
2. T40 목재 계단판
3. T1 고무판
4. T9 F.B 위 에나멜 페인트

Stair Detail

223

1. JJIMJILBANG
2. SCREEN GOLF
3. BAR-GAME ROOM
4. SUNKEN GARDEN
5. SAUNA ROOM
6. ROOM
7. LIVING ROOM
8. KITCHEN-DINING ROOM
9. WAREHOUSE
10. PORCH

1st Floor Plan

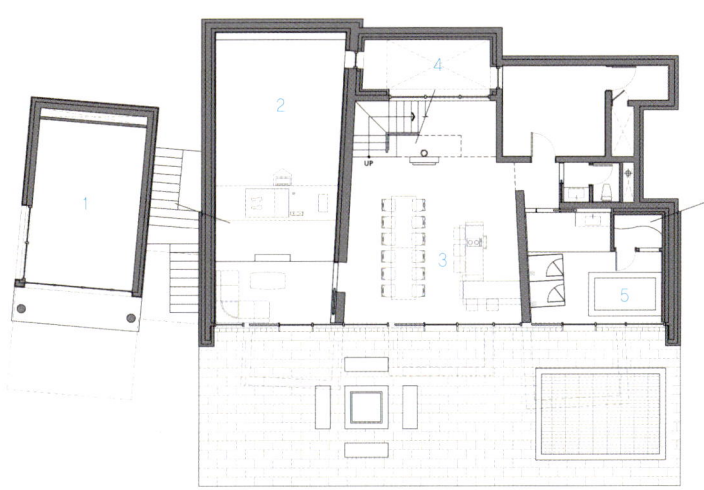

Basement 1st Floor Plan

JAEHO HOUSE Jeju, Korea

JNPeople Architects

Architects JNPeople Architects / Sehwan Jang, Seonghee In **Location** Jeju, Korea **Site Area** 1,321.3㎡ **Bldg. Area** 204.87㎡ **Gross Floor Area** 256.61㎡ **Bldg. Scale(Floor)** 2FL **Structure** Light Weight Wood Frame System, Heavy Timber Construction, Reinforced Concrete **Max. Height** 8.9m **Exterior Finish** Magic Stone **Photos Offer** JNPeople Architects

설계 제이앤피플 건축사사무소 / 장세환, 인성희 **위치** 서귀포시 남원읍 남원리 2458-4 **대지면적** 1,321.3㎡ **건축면적** 204.87㎡ **연면적** 256.61㎡ **규모** 지상2층 **구조** 경량목구조, 중목구조, 철근콘크리트조 **최고높이** 8.9m **외부마감** 매직스톤 **사진제공** 제이앤피플 건축사사무소

Site Plan

소통테 라스
부부동
자녀동
부인을 위한 물부엌 마당
소통복도
남편의
여가 활동을위한 온실 마당
한옥의 서까래를
모티브로한 캐노피
부모동
한옥의 툇마루와 서까래
부모를 위한 귤 마당
가족마당
기존창고

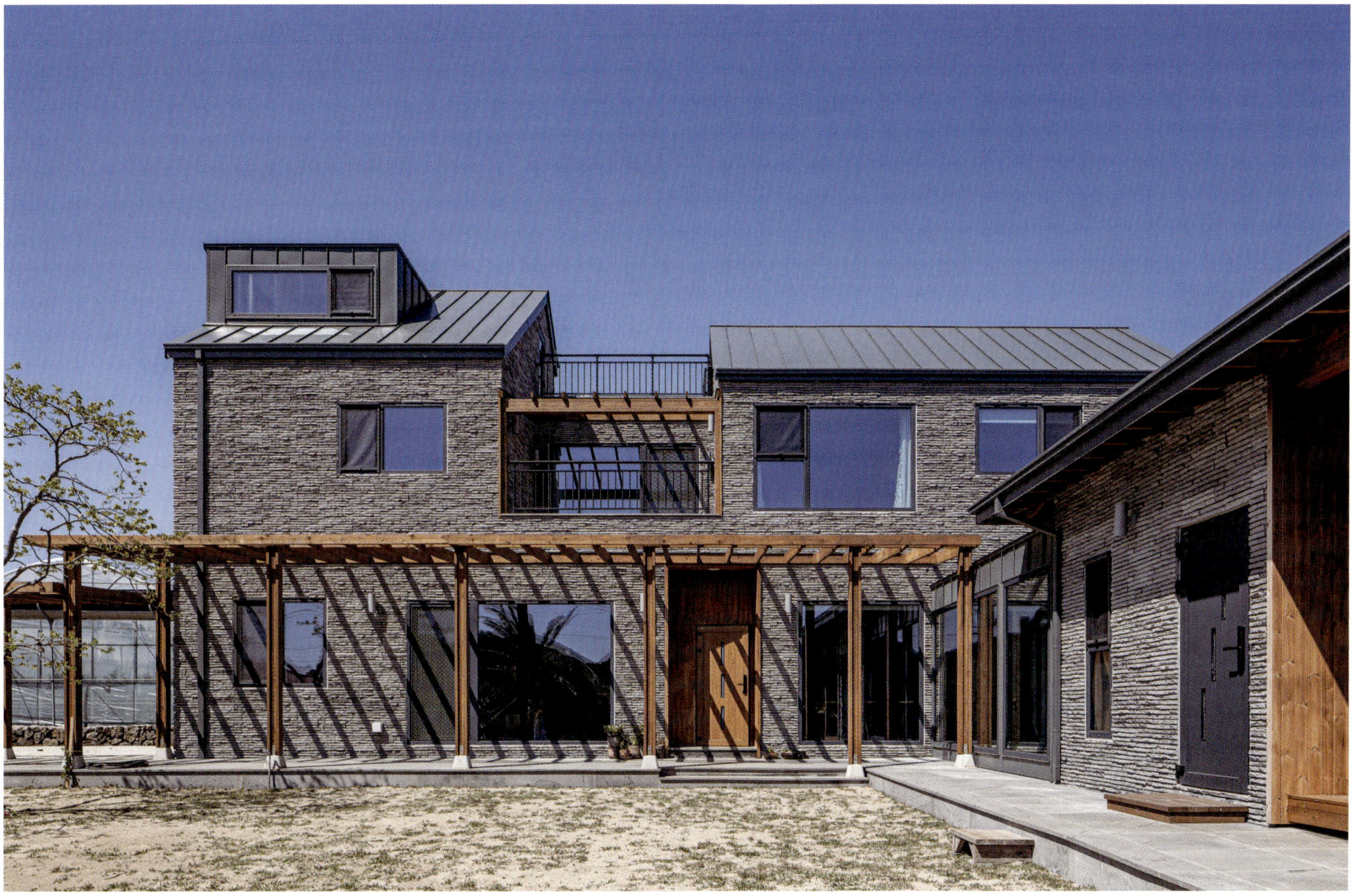

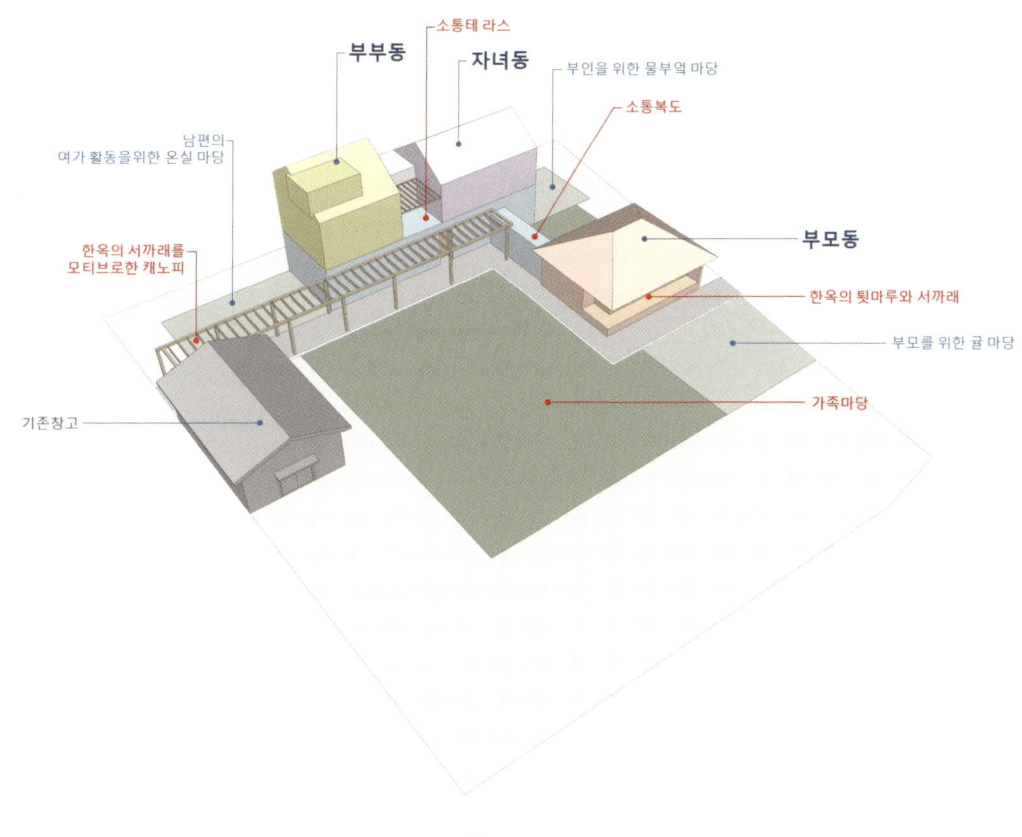

Diagram

The owner went to Seoul from Jeju Island when he was young, and now he has passed middle age and has reached the age of retirement, where he was born and raised again. We named it "Jaeho House" after the owner's first name, "Jaeho".

Things that were there
It was not only the parents who were waiting for Jae-ho, who had gone up to Seoul a long time ago. A house filled with memories, a young bija tree that I used to ride and play as a child has turned into a gloomy screen, and my father's tangerine tree, stone walls, and the road I used to walk to school… The house is old, but the trees have grown up to embrace a new family. It became a huge bosom to give. Much of it was still there before Jaeho moved to Seoul. I wanted to protect the memories that my family had made together and the many things that contained them, and I thought that a space for a family does not necessarily have to be new. In particular, the location of the warehouse building on the left side of the site that my father built himself was not bad, and the interior timber frame was quite impressive. Therefore, it was strongly recommended to the client to use it as a workshop or cafe in the future, and an arrangement plan was made based on this.

Reality
The owner, wife, father, mother, and son who can be together are undoubtedly a family, but they have all lived in different spaces for decades, doing different things. Of course, the most important thing is a space to share with each other, but now the work you like and the space you need are different. It was also essential to organize internal and external spaces for each member of the family.

The site of the old house that my father had been protecting for a long time and the basalt

Basically, after securing a comfortable living environment, I wanted to protect as much as possible the old and familiar things. The house materials were stained with hand stains, the tangerine tree that you could always see through the window when you woke up, the windowsills were lowered so that you can see them even when you are older, and a floor is provided under the rafters in the warm sunlight. However, the use of some internal materials as they are was not realized due to disagreement with the builder. So, once the location of parental consent was maintained, the location of the existing house was maintained.

The eldest son and the eldest daughter-in-law of the grandparent's house, the owner couple
The grandson of the Hyun family and the eldest daughter-in-law requested a separate ancestral hall for various family events and a water kitchen of sufficient size. In particular, it is good for a father and son to live together, but since they have lived apart from the son and wife for a long time, proper spatial separation was also necessary. Therefore, the parents' building was prepared as an annex in the existing house location and connected to the kitchen by placing a corridor for communication.

External space for each family member that contains the personality of the family
An external space for the owner of the building, which will be used as a greenhouse in the future, an external space connected to the wife's water kitchen, and a tangerine yard for the father's pastime were planned.

Son to live with
The son's space, who will live together in the future, is configured in the form of a studio on the right and east side of the second floor for privacy, and a separate entry and exit route is provided.

Front Elevation(Couple Area)

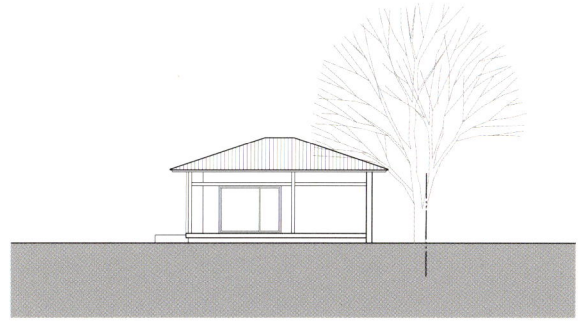

Front Elevation(Parent Area)

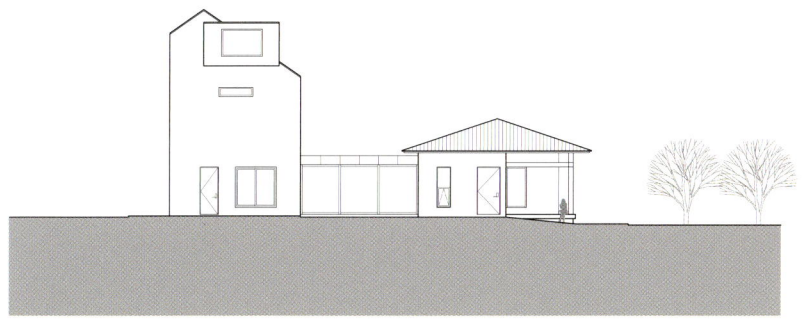

Left Elevation

Right Elevation

재호가(在好家)

건축주는 어렸을 적 제주도에서 서울로 상경하였고 이제 어느덧 중년을 넘어 은퇴의 나이가 되어 이제 다시 나고 자란 그곳, 앞으로는 서귀포 앞바다 뒤로는 한라산이 보이는 그곳으로 돌아와 지병이 있으신 어머니와 아버지와 함께 할 집을 짓고자 하였다. 건축주의 성함인 "재호"를 따서 "在好家"라는 이름을 지었다.

■ 그곳에 그대로 있던 것들

오래전 서울로 올라간 재호를 기다리고 있던 것은 부모님 뿐만이 아니다. 추억이 깃들어 있는 집, 어렸을 적 타고 놀던 어린 비자 나무는 어느덧 우람한 병풍이 되었고 아버지가 가꾸던 귤나무, 돌담, 학교를 갈 때 거닐던 길... 집은 낡았지만 나무들은 훨씬 자라 새로운 가족을 품어줄 커다란 품이 되었다. 그곳의 많은 것들은 여전히 재호가 서울로 올라가기 전 그대로였다. 이렇듯 가족이 함께 만들어 왔던 추억과 그 추억 깃들어 있던 많은 것들은 지켜드리고 싶었으며, 가족을 위한 공간이 반드시 새로운 것일 필요는 없다고 생각했다. 특히 아버님이 직접 지으신 대지 좌측의 창고 동은 그 위치가 나쁘지 않았고, 내부의 팀버프레임이 상당히 인상적이었다. 그래서 건축주에게 추후 작업장이나 카페 등으로 활용할 것을 적극 권장하였고 이를 토대로 배치 계획을 하였다.

■ 현실

건축주, 아내, 아버지, 어머니, 그리고 함께할 수도 있는 아들 가족 임에는 틀림없지만 모두 수십 년 동안 다른 공간에서 다른 것을 하며 살아왔다. 물론 제일 중요한 것은 서로 함께할 공간이지만, 이제 좋아하는 일과 필요하는 공간이 서로 다를 수 밖에 없다. 가족 각각의 구성원을 위한 내·외부공간 구성도 필수였다.

■ 아버지가 오랫동안 지켜왔던 낡은 집터와 현무암

기본적으로 쾌적한 생활환경을 확보한 뒤 최대한 기존, 시간의 때가 묻고 익숙한 많은 것들은 지켜드리고 싶었다. 손때가 묻은 집안의 자재들, 일어나면 창 너머로 늘 보였던 귤나무의 모습, 더 나이가 드시더라도 보실 수 있게 창턱은 낮추었고 따뜻한 햇볕이 드는 서까래 밑에 툇마루도 마련해 드렸다. 하지만 일부 내부 자재들을 그대로 사용하는 것은 시공자와 이견으로 실현되지는 못했다. 그래서 일단 부모님 동의 위치는 기존 주택의 위치를 유지하였다.

■ 종갓집의 맏아들 그리고 맏며느리인 건축주 부부

현씨 집안의 종손 그리고 맏며느리 각종 가족행사를 할 수 있는 별도의 제사룸과 충분한 크기의 물부엌 등을 요청하였다. 특히, 아버지와 아들이 함께 거주하는 것은 좋은 일이지만 오랫동안 아들 내외와 떨어져 살아왔기 때문에 적절한 공간적 분리도 필요했다. 그래서 부모님 동은 기존의 집 위치에 별채로 마련 소통의 복도를 두어 본채 주방쪽으로 연결하였다.

■ 가족의 개성이 담긴 가족 구성원별 외부공간

추후 온실로 활용 될 건축주분을 위한 외부공간, 부인의 물부엌과 연계된 외부공간, 아버님의 소일거리를 위한 귤나무 마당을 계획하였다.

■ 추후 함께 살게 될 아들

추후 함께 살게 될 아들의 공간은 프라이버시를 위해 2층 우측, 동측에 원룸형식으로 구성하였고 별도의 진출입 동선을 마련하였다.

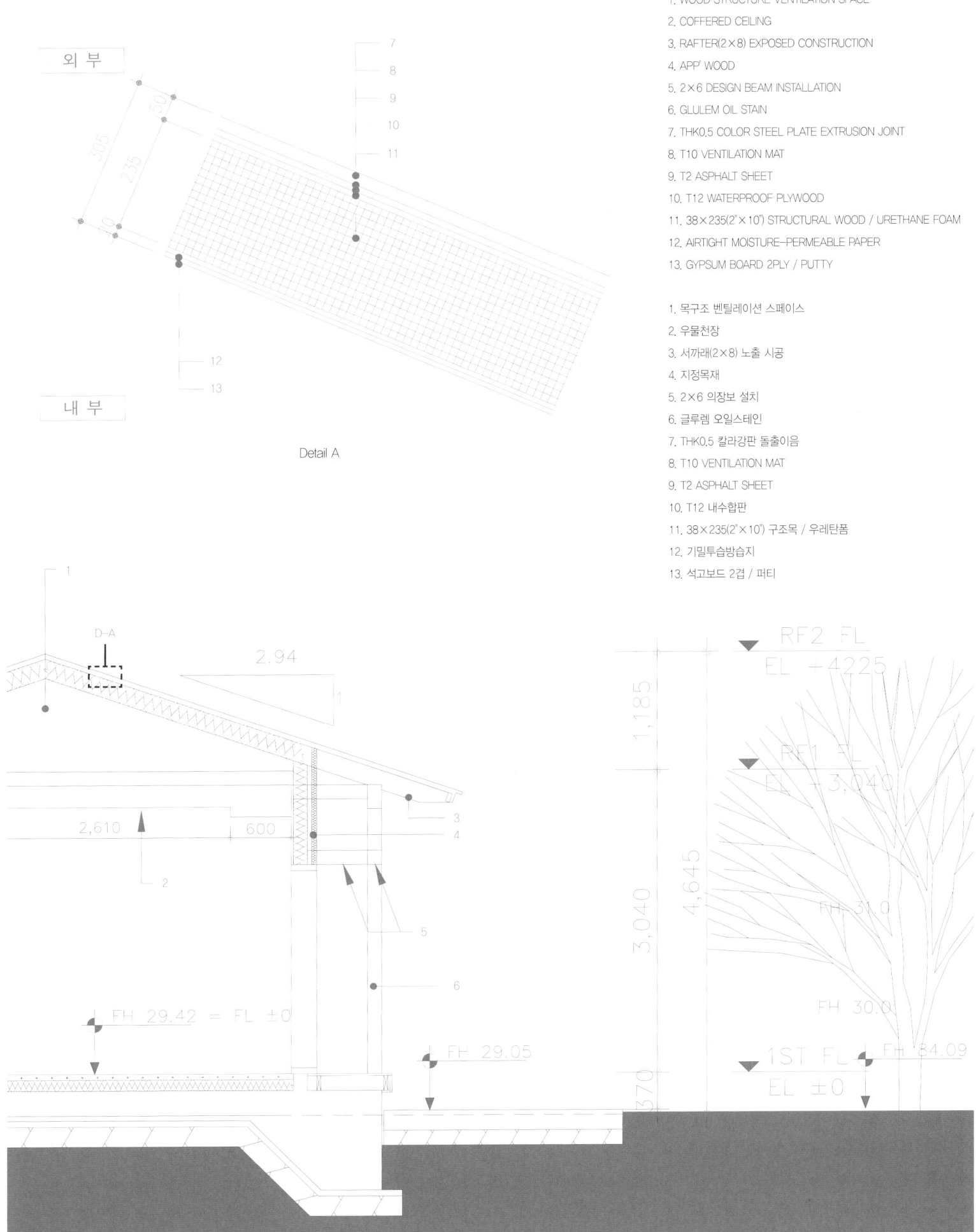

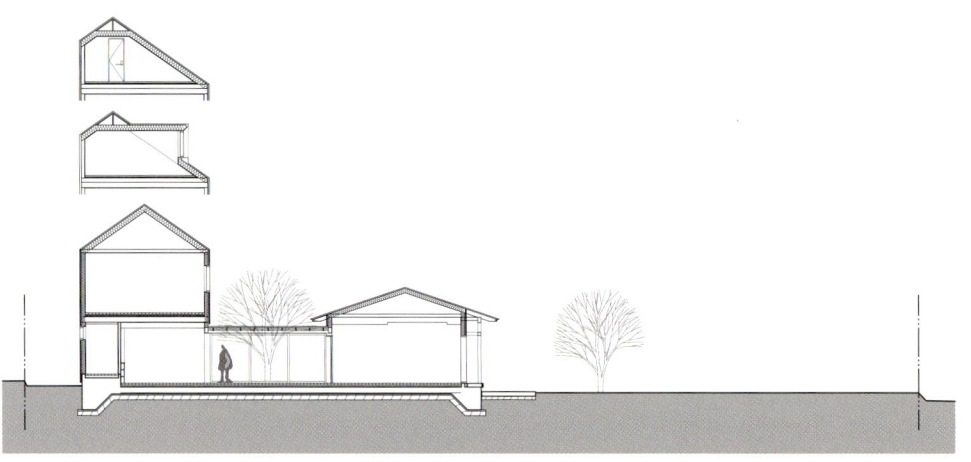

Longitudinal Section A

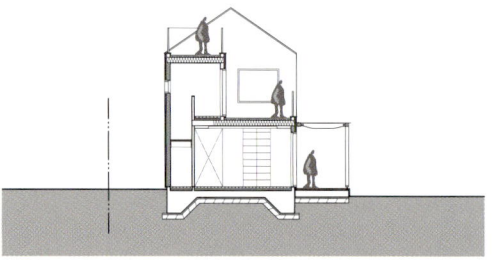

Longitudinal Section B

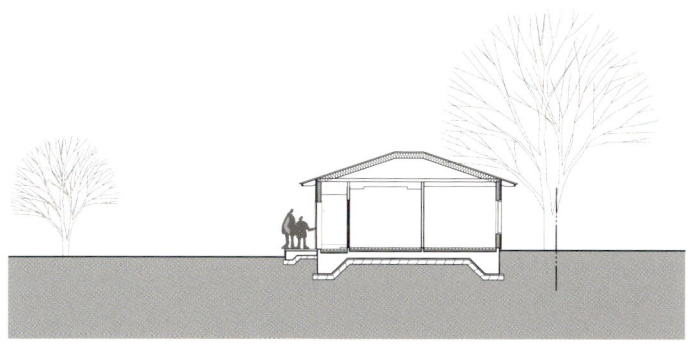
Cross Section(Parent Area)

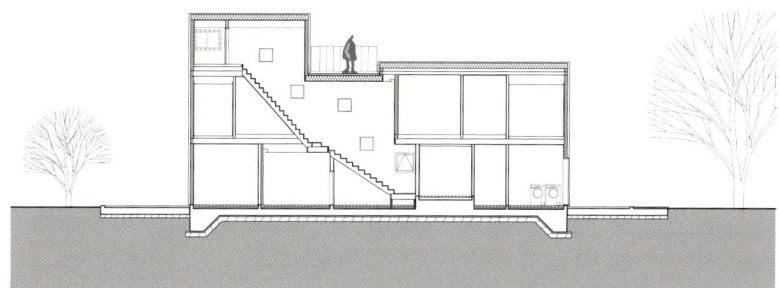
Cross Section(Couple Area)

1. AUXILIARY WAREHOUSE(EXISTING)
2. ROOM
3. LIVING ROOM
4. KITCHEN & DINING
5. MAIN ROOM

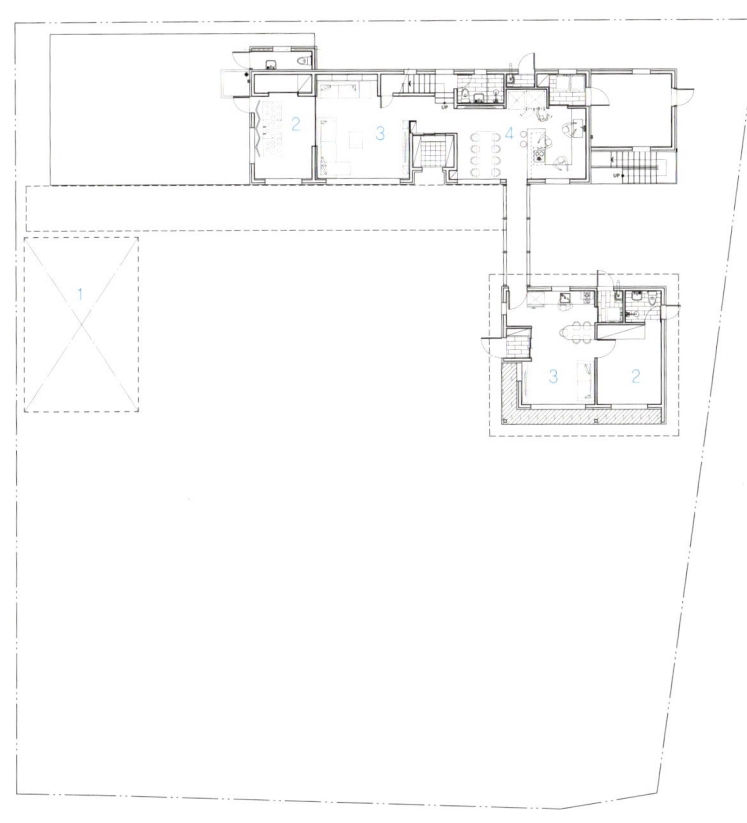

1st Floor Plan

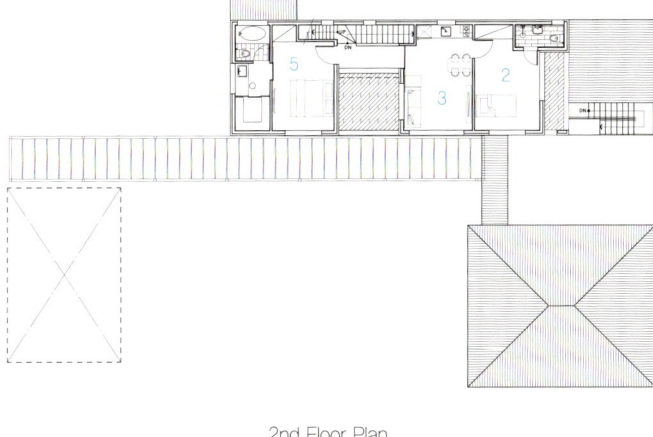

2nd Floor Plan

Bi-House Namyangju, Korea

WELLHOUSE

Architects WELLHOUSE / Jeguen Myung **Location** Namyangju, Korea **Site Area** 1,283㎡ **Bldg. Area** 222.91㎡ **Gross Floor Area** 347.33㎡ **Structure** Reinforced Concrete, Wood Structure **Exterior Finish** Stucco Paint **Photographer** Younggi Min

설계 웰하우스종합건축사사무소 / 명제근 **위치** 경기도 남양주시 수동면 입석리 **대지면적** 1,283㎡ **건축면적** 222.91㎡ **연면적** 347.33㎡ **구조** 철근콘크리트, 목구조 **외부마감** 스터코 도장 **사진** 민영기

Site Plan

Sketch

수동 바이-하우스

대지는 수동 골짜기 두목(산이 반원형으로 돌아서 생긴 마을에서 유래) 고개라는 곳에 위치하고 있다. 대지의 환경이 자연 깊숙한 곳에 자리하여 산세의 위압감과 계곡에서 불어오는 바람, 동쪽 아래에 위치한 불교 건축물과의 관계성 등을 고려하여 설계하였다. 바이-하우스의 배치는 'ㄷ'자형으로 주축은 간좌곤향(북서쪽을 등지고 남동향 조망)과 주택의 중심공간인 필로티, 2층 가족실이 위치한 축은 건좌손향(북서쪽을 등지고 남동향 조망)의 배치로 각 공간과의 유기적인 관계성을 갖도록 하였다. 바이-하우스는 현대식의 모던한 공간에 한국식 건축양식을 접목하여 구성함으로써 편리함과 한옥의 정취를 동시에 누릴 수 있도록 하였다.

The site is located in a place called Soodong Valley dumok_Boss (derived from a village formed by a semicircular mountain) pass. It was designed in consideration of the majesty of the mountains, the wind blowing from the valley, and the relationship with the Buddhist buildings located in the lower east, as the environment of the land is located deep in nature. The layout of the Bi-House is 'U'-shaped, and the main axis is Ganjwa Gonhyang (southeast view with the northwest back), and the axis where the Piloti, the central space of the house, and the family room on the 2nd floor are located, is arranged in the direction of the left hand side (southeast view with the northwest back). to have an organic relationship with Bi-House combines the Korean architectural style with a modern space so that you can enjoy the convenience and atmosphere of a hanok at the same time.

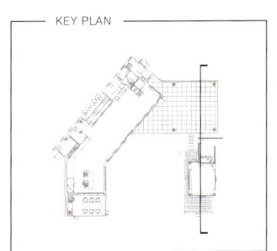

KEY PLAN

1. PILOTI
2. CHUNG-JI-KAN
3. ROOM(SARANGBANG)
4. NUMARU(UPPER FLOOR)
5. CLOSET
6. BATHROOM
7. CORRIDOR
8. BEDROOM
9. BEDROOM(ATTIC)

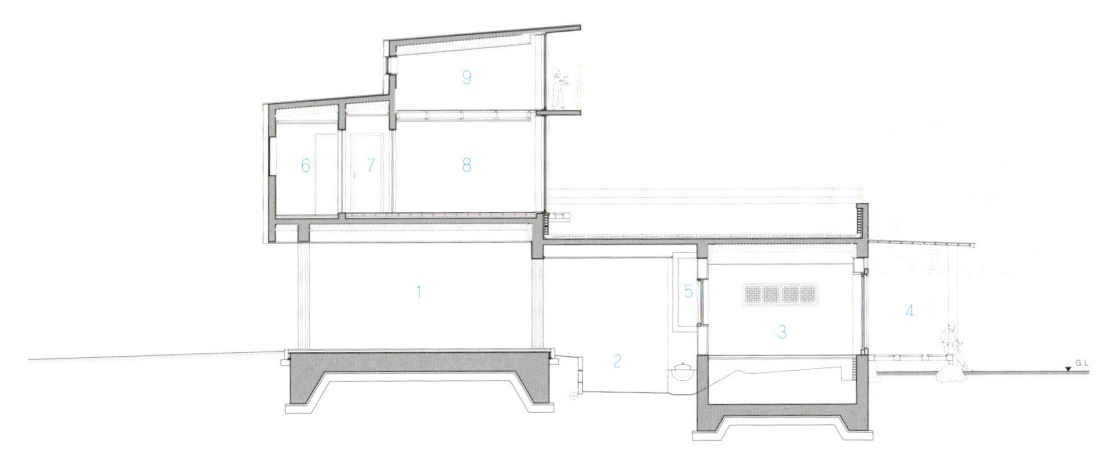

Section A

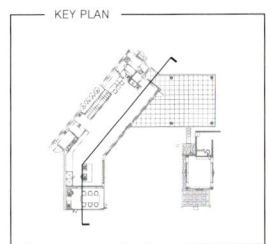

KEY PLAN

1. FRONT ROOM
2. LIVING ROOM
3. DINING ROOM & KITCHEN
4. TERRACE
5. FAMILY ROOM
6. CORRIDOR

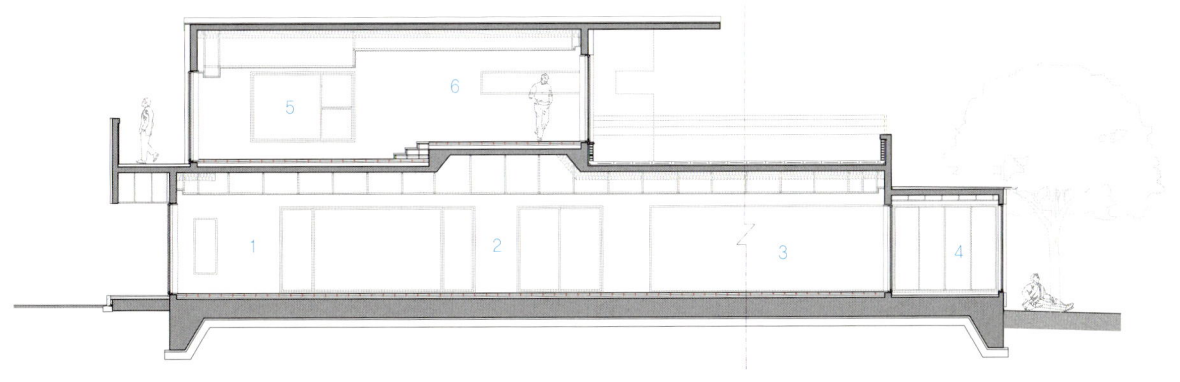

Section B

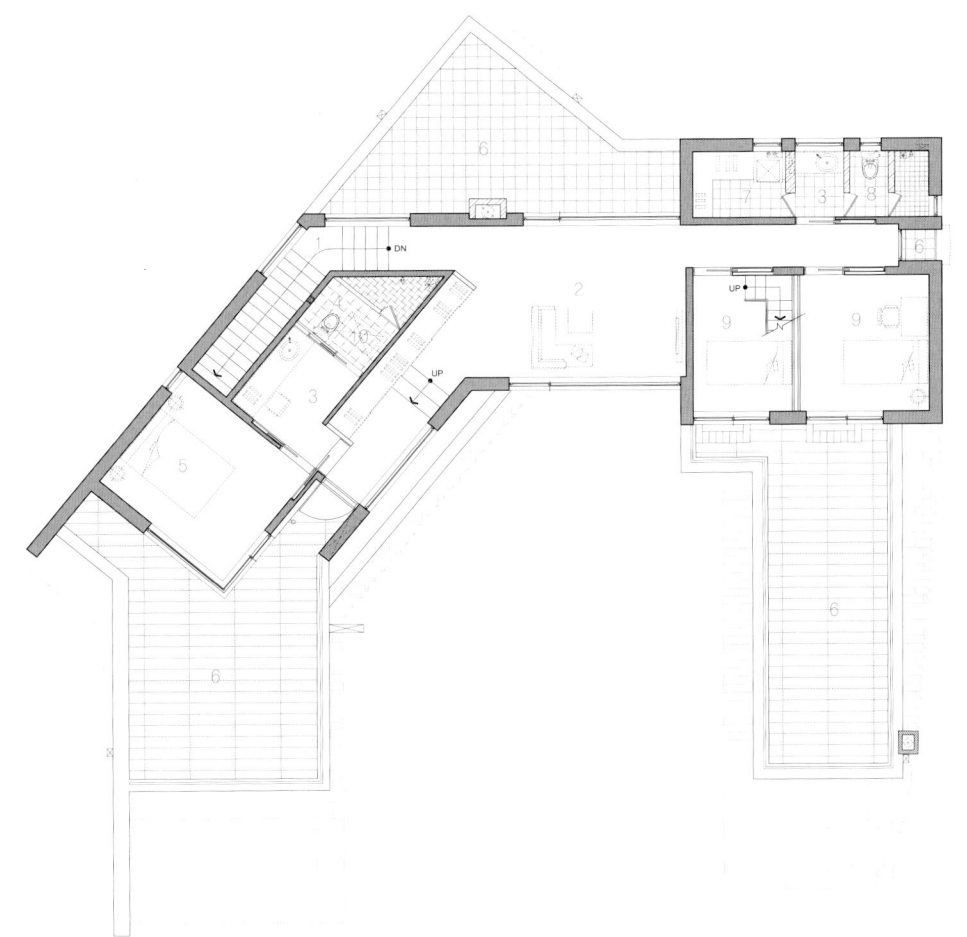

2nd Floor Plan

1. STAIRCASE
2. FAMILY ROOM
3. WASHROOM
4. COUPLE'S BATHROOM
5. COUPLE'S BEDROOM
6. TERRACE
7. LAUNDRY ROOM
8. SHARED BATHROOM
9. BEDROOM
10. SKYLIGHT

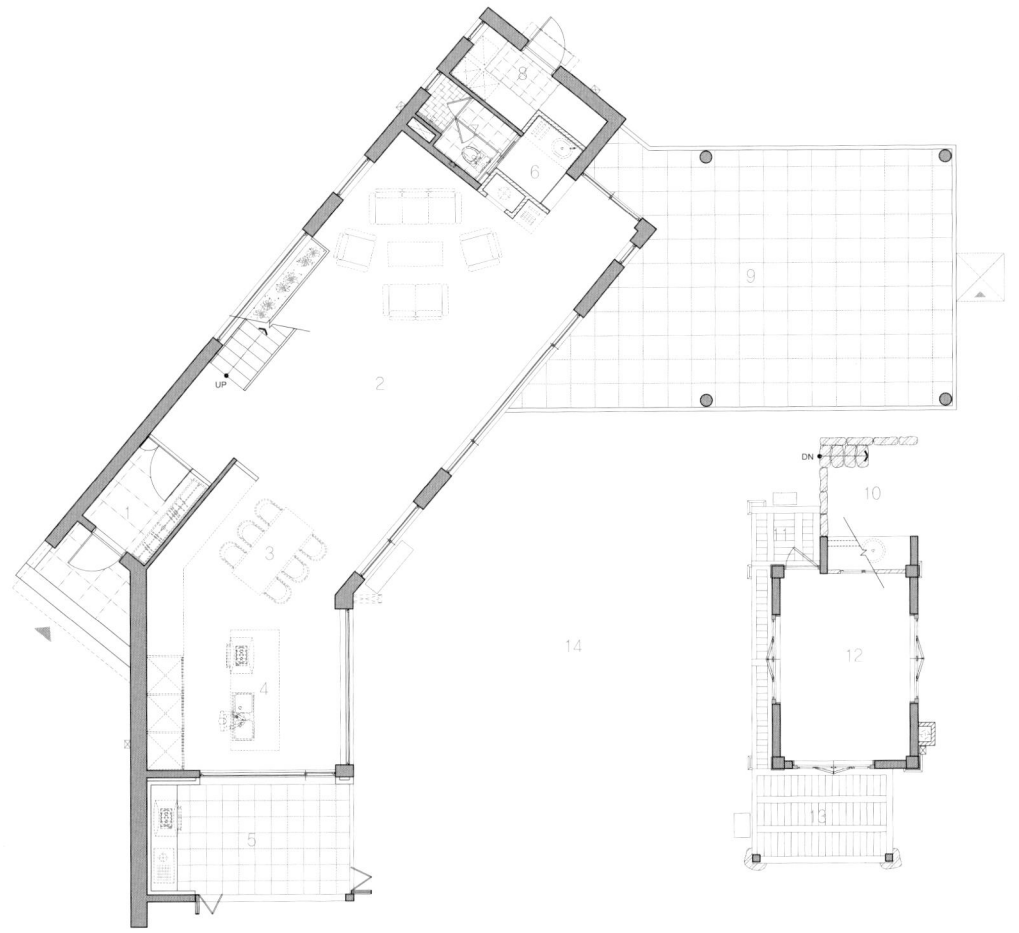

1st Floor Plan

1. PORCH
2. LIVING ROOM
3. DINING
4. KITCHEN
5. TERRACE
6. WASHROOM
7. SHARED BATHROOM
8. BOILER ROOM & WAREHOUSE
9. PILOTI
10. CHUNG-JI-KAN
11. TOENMARU
12. DETACHED HOUSE
13. UPPER FLOOR
14. COURTYARD
15. FAUCET

1. BEDROOM(ATTIC)
2. BALCONY

3rd(Attic) Floor Plan

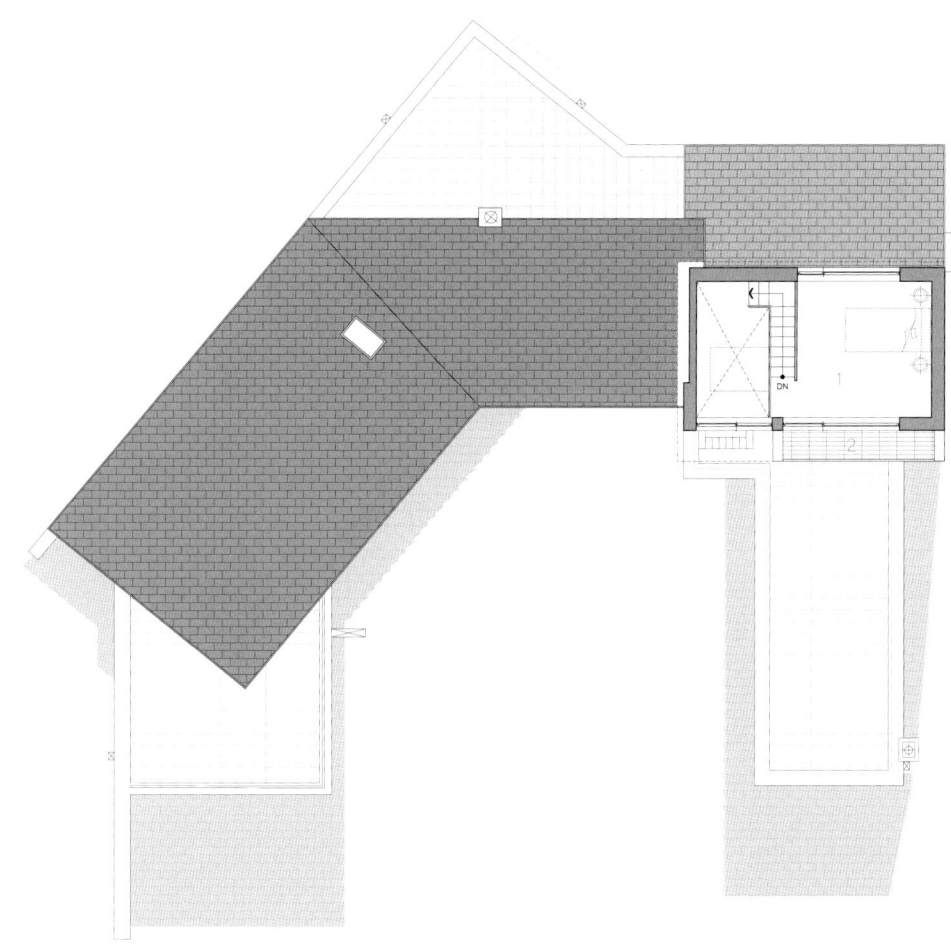

Kang An Dang Goseong, Korea

Sung Architects Group

Architects Sung Architects Group / Sunggon Kim **Project Team** Chungheon Nam **Location** Goseong, Korea **Site Area** 1,746㎡ **Bldg. Area** 225.09㎡ **Gross Floor Area** 191.81㎡ **Bldg. Scale(Floor)** 2FL **Structure** Reinforced Concrete **Max. Height** 8.2m **Exterior Finish** Staco, C-Black **Client** Jeongyoon Lee, Cheolwoo Kim **Photos Offer** Sung Architects Group

설계 성종합건축사사무소 / 김성곤 **설계참여자** 남충헌 **위치** 경상남도 고성군 동해면 조선특구로 1753-12 **대지면적** 1,746㎡ **건축면적** 225.09㎡ **연면적** 191.81㎡ **규모** 지상2층 **구조** 철근콘크리트 **최고높이** 8.2m **외부마감** 스타코, 씨블랙 버너구이 **발주처** 이정윤, 김철우 **사진제공** 성종합건축사사무소

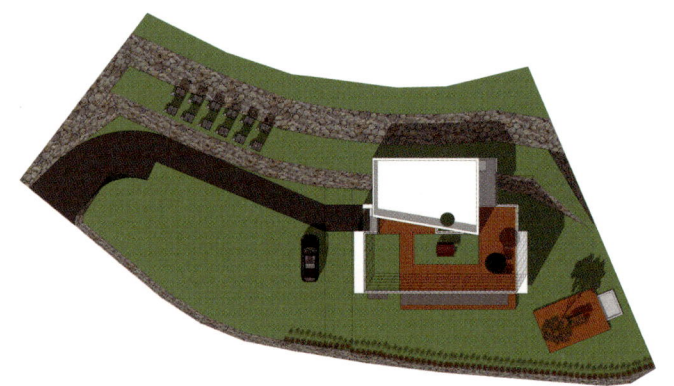

Site Plan

Elevation I

Elevationㅂ II

Sketch

247

The site is located in the middle of a mountain, and the blue sea and wide fields can be seen in front of it. The sea is always blue, but the fields change clothes depending on the season, filling and emptying them repeatedly. These scenery are good places where there is no bias between the sea and the field. As for the layout, the house sits on the lower part of the first floor and the upper part of the second floor in a cascade so that it conforms to the land developed in a narrow stepwise manner. The somewhat lacking yard in front of the living room on the 1st floor is not lacking in the view by securing a deck yard using the roof on the 1st floor.

The floor plan is external-oriented in consideration of the shape of the site, the surrounding environment, and the view, and light and nature of red maple leaves are introduced into the indoor courtyard. The common space of the living room and dining room faces the sea and field to the east with a good view, and the rooms on the first and second floors face south. There is no opening on the west side, and energy saving is considered by arranging common spaces such as stairs and a utility room on the north side. A circular skylight is installed on the ceiling of the direct staircase leading to the second floor, providing natural light by day and starlight by night.

The water-friendly space in the pond in front of the living room is a device that cools the geothermal heat in summer as well as emotional stability. When the wind blows over the calm water, a small wave forms. Just like a calm mind is shaken when worries arise... You learn wisdom from nature.

The exterior and color were designed in a simple and modern way with horizontal stability considering the terraced location. The color is a simple contrast between white and gray. In architecture, color is reduced to achromatic colors, and nature, humans, and interior accessories add color to the space and architecture.

The interior is also unified in white, excluding colors and decorations that disturb the vision when contemplating nature.

1. EXTERNAL INSULATION DRY BIT THK100 SHELL BOARD INSULATION STYROPOLE	1. 외단열 드라이비트 THK100 쉘보드 단열스치로폴
2. BEAN GRAVEL PAPER	2. 콩자갈깔지
3. STRETCH JOINT@1,500×1,500 (6×9 ASPHALT PRIMER FILLING)	3. 신축줄눈@1,500×1,500 (6×9 아스팔트 프라이머 충진)
4. THK50 MACHINE PLASTERING (REMY 1:3)	4. THK50 기계미장(레미1:3)
5. LIQUID WATERPROOFING TYPE 1	5. 액체방수 1종
6. THK180MM INSULATION STYROPOLE	6. THK180MM 단열스치로폴
7. WOOD CEILING FRAME	7. 목재천장틀
8. APP' CEILING PAPER ON 2PLY OF GYPSUM BOARD	8. 석고보드 2겹 위 지정 천장지
9. CURTAIN BOX	9. 커튼박스
10. DRY BIT (MESH) FINISH (WHITE)	10. 드라이비트(MESH) 마감(백색)
11. INSIDE CURTAIN WALL: GRAY WATER-BASED PAINT ON PLASTER	11. 커튼월 내부: 미장에 회색 수성 페인트
12. COLOR ALUMINUM SYSTEM CURTAIN WALL THK35 COLOR TRIPLE GLASS	12. 칼라알루미늄 시스템커튼월 THK35칼라 3중유리
13. APP' FLOOR BOARD	13. 지정 강마루판
14. HOT WATER ONDOL	14. 온수온돌
15. APP' WALLPAPER FIN ON GYPSUM BOARD 2PLY	15. 석고보드 2겹 위 지정벽지 마감
16. APP' WALLPAPER FIN ON GYPSUM BOARD 2PLY ON THK4MM HEAT-REFLECTING INSULATION	16. THK4MM 열반사단열재 위 석고보드 2겹 위 지정벽지마감
17. ST HANDRAIL INSTALLATION(BLACK)	17. ST난간설치(BLACK)
18. THK400MM FOUNDATION CONCRETE	18. THK400MM 기초콘크리트
19. THK140MM INSULATION MATERIAL('A' GRADE)	19. THK140MM 단열재(가등급)
20. THK60MM PLAINE CONCRETE	20. THK60MM 버림콘크리트
21. PE FILM PAPER(2PLY)	21. PE필름깔지(2겹)
22. COMPACTING THE EXISTING GROUND	22. 기존지반 다지기
23. C-BLACK BURNER FINISH ON THK40MM HEAT REFLECTIVE INSULATION	23. THK40MM 열반사단열재 위 C-BLACK 버너구이 마감
24. FILL F.H+350	24. 성토 F.H+350
25. GROUND F.H±0	25. 원지반 F.H±0
26. THK600MM FOUNDATION CONCRETE	26. THK600MM 기초콘크리트

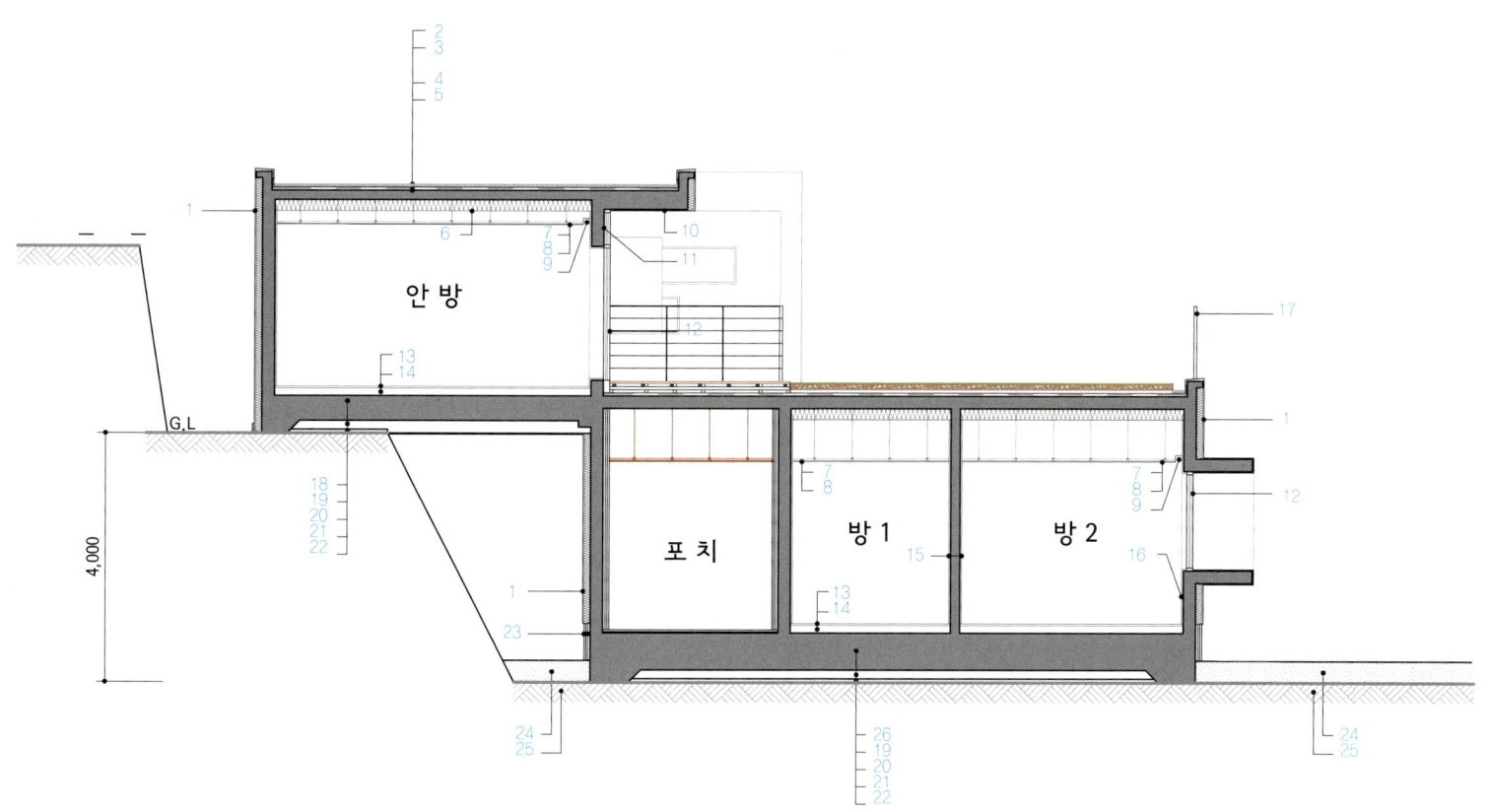

Section Detail

고성 "해품채 전원마을" (강안당)

부지는 산 중턱에 위치하며 전면으로는 푸른 바다와 넓은 들판이 한눈에 들어온다. 바다는 항상 푸르지만, 들판은 계절 따라 옷을 갈아입으며 채우고 비우기를 반복한다. 이런 풍광들은 바다와 들판 어느 한편으로 치우침이 없는 좋은 장소다.

배치는 부지 폭이 좁은 계단식으로 개발된 택지에 순응하게끔 1층은 하단부에 2층은 상단부에 계단식으로 걸쳐 집을 앉힌다. 1층 거실 앞의 다소부족한 마당은 1층 옥상을 이용한 데크마당의 확보로 전망 감상에 모자람이 없다. 열린 바다와 들판이 내 앞마당이 된다. 이 1층 옥상마당엔 잔디를 식재하고 목재데크를 설치해 친환경적이며 에너지 절감 효과도 있다.

평면은 부지의 형상과 주변 환경 그리고 전망을 고려해 외부지향형으로 하고 실내중정으로 빛과 홍단풍의자연을 들인다. 거실과 다이닝룸의 공용공간은 전망 좋은 동쪽의 바다와 들판으로 향하고, 1층과 2층의 방들은 남향이다. 서쪽으로 개구부가 없고, 북쪽으론 계단과 다용도실 등의 공용공간을 배치해 에너지 절감을 고려했다. 2층으로 오르는 직통계단 천장엔 원형 천창을 설치해 밤에는 별빛을 낮으론 자연채광을 선물한다.

거실 앞 연못의 친수공간은 정서적 안정과 함께 여름철 지열을 식혀줄 장치다. 잔잔하던 물 위로 바람이 찾아드니 작은 물결이 인다. 마치 평온하던 마음에 걱정꺼리가 생기니 마음이 흔들리는 것과 같이 자연에서 지혜를 배운다.

외관과 색상은 계단식 입지임을 고려해 수평적 안정감과 함께 심플하고 모던하게 디자인했다. 색상은 백색과 회색의 단순대비. 건축은 무채색으로 색을 줄이고 자연과 인간 그리고 실내의 소품들이 공간과 건축에 색을 더한다.

인테리어 역시, 자연을 관조하는데 시각을 어지럽히는 색상과 장식은 배제하고 백색으로 통일한다.

251

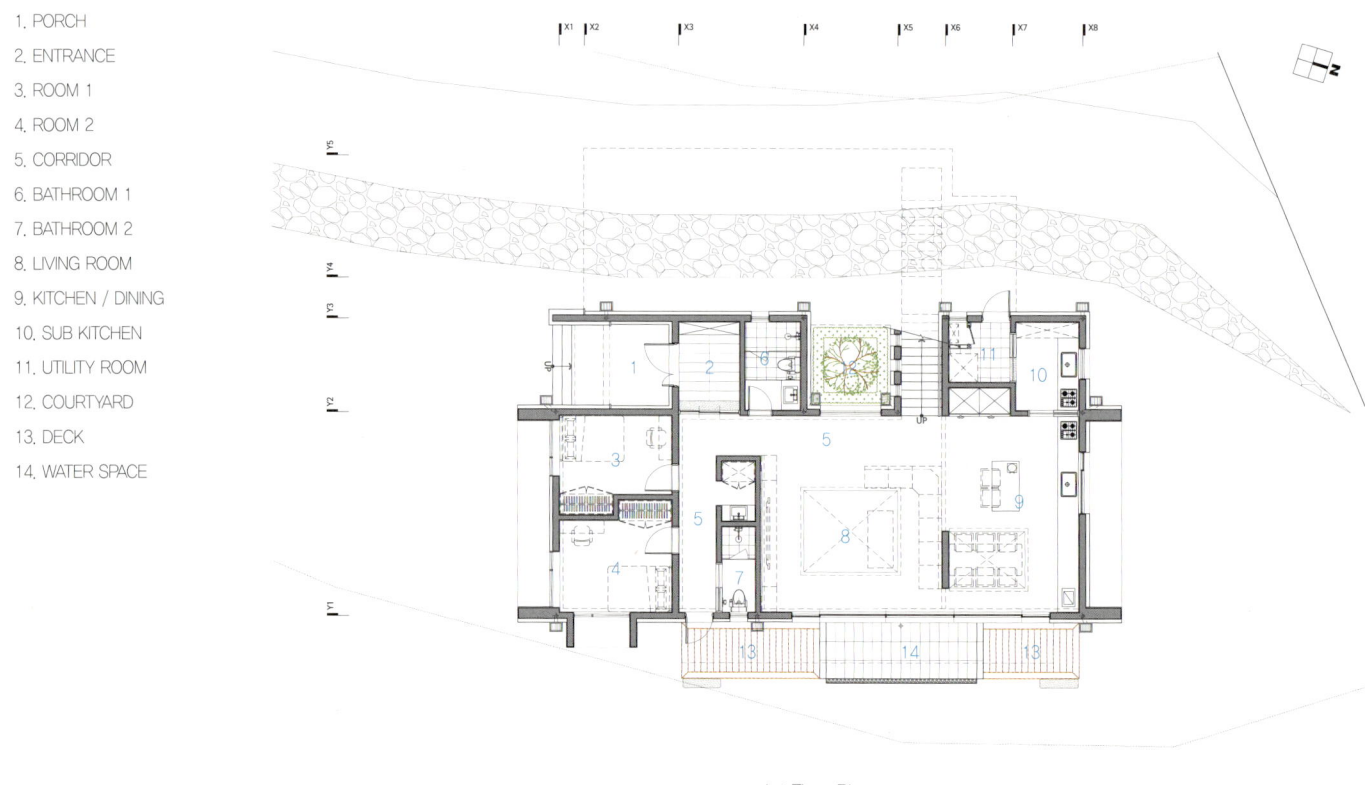

1. PORCH
2. ENTRANCE
3. ROOM 1
4. ROOM 2
5. CORRIDOR
6. BATHROOM 1
7. BATHROOM 2
8. LIVING ROOM
9. KITCHEN / DINING
10. SUB KITCHEN
11. UTILITY ROOM
12. COURTYARD
13. DECK
14. WATER SPACE

1st Floor Plan

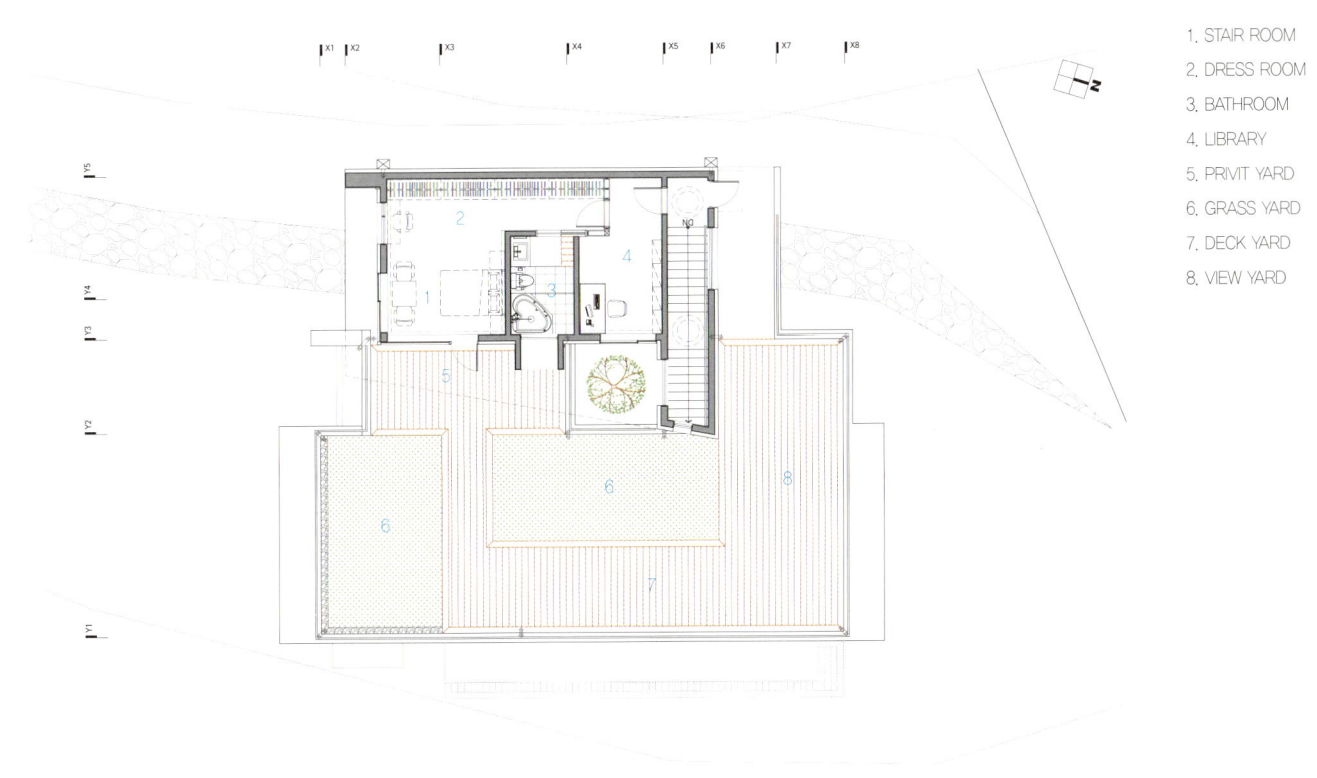

2nd Floor Plan

1. STAIR ROOM
2. DRESS ROOM
3. BATHROOM
4. LIBRARY
5. PRIVIT YARD
6. GRASS YARD
7. DECK YARD
8. VIEW YARD

Architects & Partners

One House

DRAWING WORKS
드로잉웍스

Maengdong House

How Architects Inc.
㈜하우건축사사무소

Landschaft

Architects601
아키텍츠601

Ger House

atelierjun
아뜰리에준 건축사사무소

Gaonnuri

Gong-Yu Architecture
건축사사무소 공유

Pause

Architectural Design Group HAEDAM
㈜해담건축 건축사사무소

Hokepos

NAOI + PARTNERS
나오이플러스파트너스

Oreum Madang House

JNPeople Architects
제이앤피플 건축사사무소

Cheongju Empty & Full House

RICHUE Architecture
㈜리슈건축사사무소

House_B382

Moku Architects
㈜모쿠아키텍츠

Oh U House

WE Architects
위종합건축사사무소

Cheongdo Imdang-ri House

MOON ARCHITECTS
문아키건축사사무소

HwaDam Byeol Seo

ilsangarchitects
일상건축사사무소

BB4

MaroAN Architects
마로안건축

The Sky Court House

Gyeongpiri Architecture Powerhouse
경피리건축발전소 건축사사무소

House on hill by the sea

JNPeople Architects
제이앤피플 건축사사무소

YUYUJAJEOG

admobe architect
㈜에이디모베 건축사사무소

Boryeong House

WELLHOUSE
웰하우스종합건축사사무소

Pyeongdamjae

Office for Appropriate Architecture
적정건축

Mok-dong House

JDArchitects
제이디에이건축사사무소

Green House

PLANO architects & associates
플라노건축사사무소

OYEONJAE

Zerolimits Architects
㈜제로리미츠 건축사사무소

Sangseon-won

We Architects
위종합건축사사무소

JAEHO HOUSE

JNPeople Architects
제이앤피플 건축사사무소

Bi-House

WELLHOUSE
웰하우스종합건축사사무소

Oreum Madang House

JNPeople Architects
제이앤피플 건축사사무소

Kang An Dang

Sung Architects Group
성종합건축사사무소

디자인 & 디테일
112_전원주택
Country House

발행 및 편집인 정지성
www.capress.co.kr
E-mail. capressconcept@gmail.com
capressmaru@gmail.com

기획. 취재 CA현대건축사
디자인 월간 컨셉 편집부

발행 CA현대건축사
ⓒ Copyright 2021

주소   경기 하남시 미사대로 550
현대지식산업 A10-051
대표전화 82-2-455-8043
관리 영업 82-2-455-8040
Fax. 82-031-8027-3709

인쇄 대한민국, 서울

정가 64,000원

* 이 책에 게재된 기사나 사진의 무단 복제 및 전재를 금합니다.